SpringerBriefs in Law

SpringerBriefs present concise summaries of cutting-edge research and practical applications across a wide spectrum of fields. Featuring compact volumes of 50 to 125 pages, the series covers a range of content from professional to academic. Typical topics might include:

- A timely report of state-of-the art analytical techniques
- A bridge between new research results, as published in journal articles, and a contextual literature review
- A snapshot of a hot or emerging topic
- A presentation of core concepts that students must understand in order to make independent contributions

SpringerBriefs in Law showcase emerging theory, empirical research, and practical application in Law from a global author community. SpringerBriefs are characterized by fast, global electronic dissemination, standard publishing contracts, standardized manuscript preparation and formatting guidelines, and expedited production schedules.

Alisson Alexsandro Possa

Neurorights and the Era of Neurotechnologies

An Analysis of the Brazilian Constitutional Law and the Need for New Fundamental Rights

Springer

Alisson Alexsandro Possa
Brazilian Institute of Education,
Development and Research (IDP)
Brasília, Brazil

ISSN 2192-855X ISSN 2192-8568 (electronic)
SpringerBriefs in Law
ISBN 978-3-031-92001-1 ISBN 978-3-031-92002-8 (eBook)
https://doi.org/10.1007/978-3-031-92002-8

The original submitted manuscript has been translated into English. The translation was done using artificial intelligence. A subsequent revision was performed by the author(s) to further refine the work and to ensure that the translation is appropriate concerning content and scientific correctness. It may, however, read stylistically different from a conventional translation.

This Springer imprint is published by the registered company Springer Nature Switzerland AG
The registered company address is: Gewerbestrasse 11, 6330 Cham, Switzerland

Contents

1 **Introduction** 1
References 4

2 **Neurotechnologies** 7
2.1 The Evolution of Neuroscience and Neurotechnologies 7
2.2 Perspectives of Applications of Neurotechnologies in Modern Society 10
2.2.1 Brain-Computer Interface Devices (Brain-Computer Interfaces—BCIs) 12
2.2.2 Virtual Reality Devices 14
References 15

3 **Neurorights** 17
3.1 The Legal Object Protected: Neural Data 17
3.2 New Risks 21
3.2.1 Privacy 22
3.2.2 Freedom, Integrity, and Autonomy 24
3.3 The Proposals for Neurorights 27
3.3.1 Cognitive Freedom 30
3.3.2 Mental Privacy or Neuroprivacy 31
3.3.3 Mental Integrity 32
3.3.4 Psychological Continuity 34
3.3.5 Are Neurorights New Human or Fundamental Rights? 35
3.3.6 The Trial of the First Lawsuit on Neurotechnologies: The Emotiv Case in Chile 37
References 37

4 **The Protection of Human Dignity and the Scope of Protection of Neurorights** 41
4.1 The Role of Law in Protecting Against the Risks and Damages Caused by New Technologies 41

4.2 The Protection of Human Dignity as a General Principle of Constitutional Protection 45
4.2.1 Human Dignity as an Abstract Value 46
4.2.2 The Realization of Dignity in the Face of Constitutional Practice 47
4.2.3 Human Dignity in the Brazilian Constitutional System 49
4.3 The Relationship Between Human Dignity and Fundamental Rights in the Analysis of Neurorights Proposals 54
4.3.1 The Right to Free Development of Personality and Cognitive Freedom 56
4.3.2 The Right to Physical and Psychic Integrity and Its Relations with Mental Integrity and Psychological Continuity 59
4.3.3 The Right to Privacy and Personal Data Protection and the Right to Neuroprivacy 61
References 65

5 Current State and the Future of Neurorights 67
References 70

About the Author

Alisson Alexsandro Possa holds a Master's degree in Constitutional Law from the Brazilian Institute of Education, Development and Research (IDP) in Brazil. He is affiliated with IBMEC as an Assistant Professor of History of Law and focuses his research on Brazilian constitutional law and new technologies. As a lawyer, he participated in the Brazilian Federal Senate's Bill on Artificial Intelligence, Regulation of Artificial Intelligence for the Use of the Judiciary (CNJ), and contributes to public and private institutions on the subject.

Chapter 1
Introduction

This work was developed during studies to obtain a Master's degree at the Brazilian Institute of Education, Development, and Research (IDP) in 2021, under the guidance of Professor Laura Schertel Mendes and received an Honorable Mention as one of the best dissertations in the institution's Master's program in 2022, resulting in the university's Academic Merit Recognition Award. After approval, it was updated in 2023 for publication in book format, named "Era das Neurotecnologias: O papel dos neurodireitos."[1] This English version aims to demonstrate how the topic can be approached for studies from the perspective of an already consolidated fundamental rights theory, which is the case of the Federal Constitution of Brazil of 1988, as it has more than 30 years of jurisprudential and doctrinal development.

Chile approved, on November 30, 2021, the constitutionalization of neurorights as fundamental rights, initiating the era of regulations on neurotechnologies.

This new area of technological development grows as a consequence of advances in neuroscience, a field that aggregates disciplines focused on the study of the nervous system, including biology, molecular chemistry, computing, and genetics.

While films, books, and science fiction comics explored the idea of a human being controlling electronic equipment through thoughts and interacting with virtually generated universes through devices, neuroscientific studies evolved in a way to allow such possibilities. Companies like Neuralink seek to develop equipment that enables the control of computers and mobile devices directly through neural activity.

Neurotechnologies are the methods, systems, and instruments that establish a direct connection between digital technology and the human brain, in a way that allows the recording and influence of neural activity.[2]

[1] The Era of Neurotechnologies: The Role of Neurorights (free translation).

[2] Ienca (2021), pp. 1–2.

A. A. Possa, *Neurorights and the Era of Neurotechnologies*, SpringerBriefs in Law, https://doi.org/10.1007/978-3-031-92002-8_1

These technologies, until now, were developed with a focus on health,[3] seeking new treatments for neural diseases, psychological disorders, and nervous system diseases. Currently, these innovations are being studied for other areas, such as the military, pedagogical, and entertainment,[4] foreseeing a significant increase in electronic devices that interact with brain activity, reaching the consumer market and, consequently, being used by large numbers of people around the world.

Initially, the original draft of this work addressed the name change of the company Facebook to Meta, announcing a focus on the development of a new type of virtual social space called the metaverse.[5] This new reality would be a vector for boosting neurotechnologies for entertainment purposes, as other technology companies began to announce their own versions of this strategy, including Amazon, Alphabet (Google), and Microsoft, accelerating the development of technologies to enable the realization of what was being proposed as a metaverse.

It is in this context of large investments in devices that interact directly with brain activity that concerns about their potential effects have begun to be addressed, both in the ethical and legal arenas.

Just over a year after the original draft of the book, the declining number of users on platforms that qualified as a metaverse indicated that the adoption of the idea was not as expected.[6] However, the success of the first brain implant for controlling electronic devices by Neuralink in a patient who has no body movement, allowing the control of electronic devices only with the mind[7] for a few months, until technical problems rendered the functionalities useless, demonstrates the future in the development of these products.

For the legal field, neurotechnologies bring concerns in the scope of human rights and are the subject of debate about the need for new autonomous rights.

Thus, a new proposal for normative systematization called neurorights emerged as a new human right that should be constitutionalized as a fundamental right to guarantee protection against possible damages caused by neurotechnologies. Among the main authors are Marcello Ienca, Rafael Yuste, and Roberto Andorno, whose proposals will be the subject of analysis in this book.

A terminological delimitation is necessary in this context because theories of neurorights are constructed as human rights. The terms “human rights” and “fundamental rights” are generally used as synonyms. However, the Brazilian legal doctrine differentiates between them based on positivation. That is to say, the former is present in binding international documents, while the latter is provided for in state constitutions.[8]

---

[3] Lynch (2006), p. 180.

[4] Royal Society (2019), p. 16.

[5] Meta (2021).

[6] Zitron Ed (2023).

[7] Reuters (2024).

[8] Sarlet (2018), p. 19.

The proposals for neurorights have an important element of international recognition, which will be demonstrated, which is why they are classified as human rights by the authors.

It was based on these theories that Chile approved, in 2021, the constitutionalization of neurorights in its new constitution. The understanding of Chilean congressmen regarding the nature of fundamental rights was based on the premise that the new risks to human dignity that neurodevices offer go beyond current protections, as the risks are mental and not just physical. Thus, the conclusion is that, for effective protection of Chilean citizens, the proposed international ethical standards should be consolidated as new human rights at the international level and fundamental at the national level.

In June 2023, the Supreme Court of Chile judged the first case based on the constitutional provision about neurorights, in a case against the company Emotiv,[9] showing the first practical concerns with the legal protection of the use of new technologies.

There are criticisms of the approach of neurorights as new human rights, mainly with the argument that the inflation of new types of rights with this classification can lead to the loss of meaning of their content and their importance. These criticisms are anticipated by Andorno R, Ienca M (2017) in the article "*Towards new human rights in the age of neuroscience and neurotechnology*" and are dismissed with the adoption of the criteria for new human rights proposed by Philip Alston (1985).

However, the discussion is far from being settled. In February 2022, Christoph Bublitz, who in 2013 was responsible for structuring cognitive freedom as a normative manifestation, published the article "*Novel Neurorights: From Nonsense to Substance,*" in the journal *Neuroethics*, reinforcing criticisms of current claims that all kinds of neurorights must necessarily be recognized as new human rights.

An important point of the criticism made by this author, notably in his construction that neurorights are scientific activism, is that the proposals are not based on legal theories,[10] being essential to carry out analyses from the perspective of legal sciences, so that the nature of these rights is better substantiated.

The complexity of the topic was the subject of a draft *resolution* after the 51st meeting of the Human Rights Council of the United Nations General Assembly,[11] which found the need for in-depth studies on the functioning and impacts of neurotechnologies.

Considering the concern of the risks that neurotechnologies offer to various dimensions of the human being, the parameters of analysis on the nature of these proposals will be those that are part of the construction of the defense of human dignity in Brazil, in order to observe how their characteristics can be framed in a current constitutional order.

---

[9] See Chile Supreme Court, rol n° 217225-2023 ("Girardi v Emotiv Inc").

[10] Bublitz (2022), p. 5.

[11] United Nations (2022).

It is in this factual context, where the international scenario debates proposals for rights in abstract, that this work is inserted by trying to reason at the constitutional level about the need for new dogmatic constructions of fundamental rights.

The general objective of this book, therefore, is to identify the main lines of definition of neurorights in the theories currently proposed and to carry out a dogmatic exercise to identify whether this scope of protection is already possible in a logic of protection of human dignity.

The work initially addresses an overview of the evolution of neuroscience and neurotechnologies, effectively demonstrating that there are current challenges that these developments pose for contemporary legal constructions and that the prospects for mass use in the future should be a current concern.

The second chapter analyzes the construction of the concept of legal object to be protected by neurorights, the new risks that neurotechnologies bring to this object, and a review of the main proposals of these rights in recent decades. The proposal that will be analyzed in greater detail, because it is considered the "only one that is not based on merely ethical discussions"[12] and rather on analyses of legal sciences, is the theoretical construction of Marcello Ienca and Roberto Andorno in 2017 on four new human rights published in the journal *Life Sciences, Society and Policy* (2017): (1) cognitive freedom, (2) mental privacy, (3) mental integrity and (4) psychological continuity.

Finally, the third chapter carries out an analysis of the relationship between the proposed neurorights and existing constitutional protections, starting from a construction of the protection of human dignity and its concretizations: right to free development of personality, right to physical and psychic integrity, and right to privacy and protection of personal data. As a theoretical approach to human dignity and its concretizations, the literature of Ingo Wolfgang Sarlet and Daniel Sarmento will be used together, observing positions of other authors when pertinent.

References

Andorno R, Ienca M (2017) Towards new human rights in the age of neuroscience and neurotechnology. Life Sci Soc Policy 1:27

Bublitz J (2022) Novel neurorights: from nonsense to substance. Neuroethics 1:15

Ienca M (2021) On neurorights. Front Hum Neurosci 15:701258. https://doi.org/10.3389/fnhum.2021.701258

Lynch Z (2006) Neuropolicy (2005-2035): converging technologies enable neurotechnology. Creating new ethical dilemmas. In: Bainbridge W, Roco M (eds) Managing nano-bio-info-cogno innovations, 1st edn. Springer, Dordrecht

Meta (2021) Introducing Meta: a social technology company. https://about.fb.com/news/2021/10/facebook-company-is-now-meta/. Accessed 06 Nov 2024

[12] Bublitz (2022), p. 5, footnote 11.

Reuters (2024) Musk's Neuralink to launch feasibility trial with brain implant, robotic arm. Available in: https://www.reuters.com/business/healthcare-pharmaceuticals/musks-neuralink-launch-new-feasibility-trial-with-brain-implant-2024-11-25/

Sarlet IW (2018) A eficácia dos direitos fundamentais: uma teoria geral dos direitos fundamentais na perspectiva constitucional. Livraria do Advogado, Porto Alegre

The Royal Society (2019) iHuman: blurring lines between mind and machine. https://royalsociety.org/-/media/policy/projects/ihuman/report-neural-interfaces.pdf. Accessed 06 Nov 2024

United Nations (2022) Draft Resolution A/HRC/51/L.3. https://documents-dds-ny.un.org/doc/UNDOC/LTD/G22/506/14/PDF/G2250614.pdf?OpenElement. Accessed 06 Nov 2024

Zitron Ed (2023). RIP Metaverse: An obituary for the latest fad to join the tech graveyard. Business Insider. https://www.businessinsider.com/metaverse-dead-obituary-facebook-mark-zuckerberg-tech-fad-ai-chatgpt-2023-5

Chapter 2
Neurotechnologies

This chapter provides an overview of what neurotechnologies are and explains concepts such as brain-computer interface devices (BCIs), as well as the different existing classifications. It also presents figures of the investments made in recent years for the development of these technologies in various areas, demonstrating the need and urgency for discussions about legal protections.

2.1 The Evolution of Neuroscience and Neurotechnologies

The study of the human brain dates back millennia. In Ancient Egypt, the brain was already identified as an organ of great importance, but its function remained a mystery for centuries. Only with the development of modern scientific methods, such as anatomical dissections in the Renaissance, did the first hypotheses about its role in controlling thoughts, emotions, and movements begin to emerge. This initial milestone opened doors for increasingly complex explorations.

It was in the twentieth century that neuroscience truly flourished as a robust scientific field. During this period, the foundations for what we now know as modern neuroscience were laid. The invention of the electroencephalogram (EEG),[1] which allowed the recording of the brain's electrical activities, was a watershed. Discovered in the 1920s by German psychiatrist Hans Berger, this technology enabled the observation of neural patterns in real time. Decades later, in the 1990s, considered the "Decade of the Brain" by global scientific initiatives, advances in understanding the molecular and cellular processes that shape thoughts and emotions came to light.[2]

[1] See Andorno and Ienca (2017), p. 3.

[2] See Sententia (2004), p. 223.

A. A. Possa, *Neurorights and the Era of Neurotechnologies*, SpringerBriefs in Law, https://doi.org/10.1007/978-3-031-92002-8_2

As the term "neuroscience" became established, interest in applying this knowledge to improve human life gave rise to neurotechnology. This new field combined neuroscience with cutting-edge technology to develop devices capable of interacting directly with the nervous system. Brain-computer interfaces (BCIs) emerged as protagonists of this new era, allowing brain signals to be converted into commands to operate external devices. Grey Walter, in 1964, conducted an innovative experiment by implanting electrodes in a patient's brain, showing that he could control devices just with thought. This remarkable event was an early sample of the transformative potential of neurotechnology.[3]

At the beginning of the twenty-first century, neurotechnology presented a possible new era of economic development based on brain devices.[4] This forecast shows signs of being verified with research on the amounts invested in the development of these technologies.

According to the report, *The Market for Neurotechnology 2018–2022*, in 2018 the total market was 4.8 billion dollars, reaching 13.3 billion dollars in 2022. For military uses, the *Defense Advanced Research Projects Agency* of the United States (DARPA) had a budget forecast of 100 million dollars for a biomedical technologies program. In the private sector, a survey of 250 startups focused on neuroscience showed an average market value of 4 million dollars.[5]

The Information Commissioner's Office (ICO), the UK's data protection regulatory authority, identified the values in public investments as shown in Fig. 2.1.

The numbers corroborate the predictions of the studies already mentioned, demonstrating that there is great interest on the part of states in technologies that interact directly with the brain.

Companies, like the already mentioned Neuralink, emerged with an exclusive focus on research and development in the area. In addition, large corporations such as Facebook, Amazon, and Alphabet (Google) announced research projects, attracting more and more investments since then.[6]

In a context of international competitiveness, the ten countries with the highest numbers of companies focusing on the development of neurotechnologies[7] are: United States (386), United Kingdom (39), Canada (34), Germany (22), Israel (20), Switzerland (19), France (17), Italy (14), Australia (12), and Spain (10). As for the

---

[3] See Liv (2021), p. 330.

[4] Lynch (2006), p. 229: "Since the time of the Industrial Revolution there has been a relatively consistent pattern of 50-year waves of techno-economic change. We are currently nearing the end of the fifth wave of information technology diffusion, while a sixth wave is emerging with converging advancements across the NBIC (nano-bio-info-cogno) space, making possible neurotechnology, the set of tools that can influence the human central nervous system, especially the brain."

[5] See The Royal Society (2019), p. 52.

[6] See Naysmith (2023).

[7] See Information Commissioner's Office (2023), p. 26.

Country	Value of Public Investment	Duration of the investment	Annually (to nearest million)	Investment as a percentage of GDP[23]
USA	$3.3 billion	2013 to 2023	$300 million	0.0014
EU	$1.08 billion	2013 to 2023	$98 million	0.0006
Korea, Republic of	$42 million	2021	$42 million	0.0025
Australia	$350 million	2016 to 2020	$70 million	0.0050
Japan	$307 million	2014 to 2024	$28 million	0.0005
Canada	$171 million	2011 to 2019	$19 million	0.0011
UK	$111 million	2011 to 2020	$11 million	0.0004

Fig. 2.1 Investments in research on neurotechnologies by the public sector [Source: Information Commissioner's Office (2023), p. 25]

number of patents granted for neurotechnologies by countries based on per capita, to enable the comparison of countries with different populations, it also indicates a considerable evolution since 2013 (see Fig. 2.2).

An example of a recent patent by a private company is the technology "*Biosignal Sensing Device Using Dynamic Selection of Electrodes*" from Apple,[8] which implements electrodes in the popular headphone *AirPods* that, through EEG reading, will allow the reading of the user's brain activity. It was granted to the company by the responsible body in the United States on July 20, 2023.

The current scenario allows us to verify that the public and private sectors are investing in research to expand the scope of human interaction with electronic devices: the trend of the Internet of Things, where all devices will be connected to the Internet, is already a reality, with the future being the connection with the human brain itself.

Thus, it is in future perspectives that the risks of damage are mentioned as concerns that may demand normative constructions already in the present.

[8] It is possible to find the document that granted the patent with figures illustrating how this is implemented in the company's audio playback device. See at https://image-ppubs.uspto.gov/dirsearch-public/print/downloadPdf/20230225659 (accessed on 07.11.2024).

Country	2013	2014	2015	2016	2017	2018	2019	2020	2021	2022	Total
United States	2,709	3,085	3,214	3,765	4,176	4,450	5,916	6,345	6,566	6,661	46,887
Germany	114	139	148	177	235	280	324	343	340	323	2,423
Canada	156	143	135	191	192	205	299	316	327	341	2,305
United Kingdom	111	114	94	128	139	143	217	264	291	279	1,780
Korea, Republic of	20	56	91	124	158	206	248	307	316	398	1,924
Japan	117	141	136	139	173	171	259	277	329	289	2,013
China	24	36	37	72	102	105	176	222	250	342	1,366
Israel	109	107	139	125	154	197	192	220	262	303	1,808
Switzerland	39	53	53	71	87	108	128	154	156	175	1,024
Netherlands	47	87	63	80	86	111	136	159	131	163	1,063
Rest of the world	45	68	91	67	105	138	123	122	160	76	995
Total	3,491	4,029	4,201	4,939	5,607	6,114	8,018	8,729	9,128	9,350	63,606

Fig. 2.2 Patents granted by countries to inventions of neurotechnologies based on per capita [Source: Information Commissioner's Office (2023), p. 27]

2.2 Perspectives of Applications of Neurotechnologies in Modern Society

As demonstrated, neurotechnologies are being considered the next natural step in technological evolution. Perspectives of services and products based on interaction with brain activity are already beginning to be developed in conjunction with other technological advances, such as: Internet of Things, 5G Internet for mobile devices, cloud computing and *big data*, and Artificial Intelligence for mass data analysis.

The possible uses of these technologies alongside human brain activity are conditioned to the varied modes of operation for reading and/or manipulating electromagnetic signals, which will be explained below.

First, it is necessary to describe the two main categories of neurotechnologies that may reach the market: invasive and non-invasive. The main criterion of differentiation is the location of the sensors that monitor brain activity, which can penetrate the skin or not.[9]

[9] See Steryl et al. (2016), p. 393.

This is related to the division between the central and peripheral nervous systems. According to current medical literature, the nervous system aims to make the human being relate to the environment in which they live. It is divided, basically, into:[10] central, where we have the brain (which is responsible for receiving and interpreting sensations, controlling movements, regulating vital functions, consciousness, etc.) and the spinal cord (which conducts information from the skin, joints and muscles to the brain and vice versa); and peripheral, where we find the nerves (which innervate the skin, joints, muscles (sensory and motor), internal organs, vessels and glands). That is: while the central nervous system receives, interprets, and transmits information to the other parts of the body, the peripheral nervous system ensures the performance of the action.

The main characteristic of invasive devices is that they must be inserted inside the individual's body to interact directly with neural impulses. Examples are the device for electrocorticography[11] (ECog), which operates through the insertion of electrodes on the surface of the brain to monitor electrical signals from the organ, and the cortical implant,[12] which consists of probes implanted in the cerebral cortex to record and reproduce stimuli from neurons related to muscle control, aiming to allow the control of devices solely by means of thought.

Non-invasive devices, on the other hand, are aimed at reading neural activities and use EEG (electroencephalogram), MEG (magnetoencephalography), and fMRI (functional magnetic resonance) techniques. All seek to enable the reading of brain images without the need for brain implants.[13]

In addition to being invasive or non-invasive, neurotechnologies can also be stimulating,[14] according to their ability to allow devices to stimulate certain areas of the brain to modulate, through electrical impulses, brain activities. A pertinent example is the cochlear implant,[15] which translates sounds into electrical signals. These signals then stimulate the cells of the cochlear nerve, thereby conferring the ability to listen to individuals with severe hearing deficiencies.

These types of neurotechnologies can make up two types of products: brain-computer interface devices (BCIs) and augmented virtual reality devices.

[10] See Bear et al. (2017), pp. 183–185.

[11] See Shenoy et al. (2007), p. 273.

[12] See Homer et al. (2013), p. 384.

[13] See Steryl et al. (2016), p. 394.

[14] See Andorno and Ienca (2017), p. 3.

[15] See Bowditch (2022).

2.2.1 *Brain-Computer Interface Devices (Brain-Computer Interfaces—BCIs)*

The term "brain-computer interfaces" (BCIs) was initially used to refer to both physical devices (i.e., hardware) and systems (i.e., software) that facilitate communication between humans and their environment without the involvement of the peripheral nervous system. These interfaces establish a non-muscular communication channel between the user and external devices.[16]

The operation of BCIs comprises five stages:[17] signal acquisition, signal processing or enhancement, extraction, classification, and interface control. The signal acquisition stage captures the brain signals, performing, when necessary, noise reduction and artifact treatment. In pre-processing, these signals are prepared for subsequent processing, ensuring a suitable structure for analysis. Then, the feature extraction phase identifies relevant and discriminative information present in the captured signals. The signals are then transformed into vectors that contain the essential and representative features of the recorded patterns. This process, however, is challenging, as the brain signals often mix with other signals resulting from overlapping brain activities in time and space. In addition, the signals are not necessarily stationary and can be distorted by interferences such as electromyography (EMG) and electrooculography (EOG). The resulting vector needs to be of low dimensionality to reduce the complexity of processing, without sacrificing important information. The classification stage uses the feature vectors to categorize the signals, ensuring efficient recognition of the patterns necessary to interpret the user's intentions. Therefore, the choice of quality discriminative features is fundamental for the effectiveness of this recognition. Finally, the control interface translates the classified signals into commands that can be understood and executed by connected devices, such as wheelchairs or computers, transforming brain activity into practical and functional actions.

Many of the brain signals that are translated into cognitive commands are unknown, but the advancement of studies in neurotechnology has allowed the mapping of some of them and the proper decoding of their operation to be modulated to interpret intentions and translate them into commands.

These devices can be classified into the following categories: active BCIs, reactive BCIs, and passive BCIs.[18] Active BCIs are those whose commands are issued to devices consciously and intentionally by the user, with the technique of *motor imagery* being the most known. This technique consists of the user's activity of imagining a physical action so that the device reproduces the action. Reactive BCIs are those whose commands are derived from the interaction of external stimuli with the user's brain activity. They are being used to control Internet browsers and even to create paintings derived from brain activity. Finally, passive BCIs are those that

[16] See Gomez-Gil and Nicolas-Alonso (2012), p. 1212.

[17] See Gomez-Gil and Nicolas-Alonso (2012), p. 1212.

[18] See Liv (2021), p. 331.

detect neural stimuli and perform actions without the need for will or any user interaction, allowing a real-time analysis of signals that translate into emotional and mental states. These BCIs are being evaluated to perform evaluations of abilities, vehicle driving, among other utilities.

The different types of BCIs are being researched and developed in various areas, which will be exemplified below.

As initially exposed, the main area of development of neurotechnology up to the present moment was the medical one. The applications of BCIs in the medical context[19] are useful palliative measures for neurological diseases associated with the gradual loss of sensory and motor functions, total disabilities of motor functions, and impossibility of locomotion

The Royal Society[20] highlights, among the applications pointed out, cochlear implants or *deep brain stimulation* (DBS), used to treat Parkinson's disease; the technique of Functional Electrical Stimulation (FES), to assist individuals in recovering motor function, and the technique of Transcutaneous Nerve Electrical Stimulation (TENS), used to relieve pain. The medical area and the well-being sector will be the most impacted by the use of neurotechnologies for the next 3 years.

In this short-term scenario, the use of neurotechnologies is also predicted to gain strength for professional sports through non-invasive neurotechnologies in athletes, to measure responses to stimuli and concentration levels.

The field of education is one in which neurotechnologies may have a significant impact. China has emerged as a notable exemplar in this regard, with a significant focus on research endeavors exploring the potential of utilizing external devices to assess levels of concentration and stress. Additionally, there is a growing interest in the prospect of neuromodulation of cognitive processes to enhance academic performance.[21] Another example from the country are projects of neurotechnologies for monitoring productive capacity in work environments.[22]

Finally, the main use of BCIs that will probably lead to mass consumption of neurotechnologies is in entertainment.[23]

As already mentioned, private company investments for the development of neurodevices have already reached hundreds of millions of dollars per year. Many of these companies are geared towards mass consumption, mainly in entertainment and communication.

---

[19] See Gomez-Gil and Nicolas-Alonso (2012), p. 1257.

[20] See The Royal Society (2019), p. 20.

[21] The Royal Society (2019), p. 101, note 224.

[22] See Liv (2021), p. 333.

[23] See Andorno and Ienca (2017), p. 5: "Portable, easy-to-use, consumer based transcranial direct current stimulation (tDCS) devices are the most widespread form of consumer-grade neurostimulator. They are used in a number of low-cost direct-to-consumer applications aimed at optimizing brain performance on a variety of cognitive tasks, depending on the brain region being stimulated."

The companies Emotiv[24] and Neurosky[25] already offer wireless headsets that allow control of smartphones and computers with thought alone.

It is in this context of entertainment that a second category of neural devices is emerging for the mass consumer market: devices for virtual reality.

2.2.2 *Virtual Reality Devices*

The technology equipment market has been providing devices for virtual reality since 2016: Facebook, with its Oculus Rift, Sony, with PlayStation VR, and Valve, with The Vive. These devices did not originally interact with brain activity, serving only as audio and image players.

It happens that new technologies that blend virtual reality (VR) and augmented reality (AR) devices are emerging under the name of extended realities (XR), like the case of Apple Vision Pro, which already allows the incorporation of accessories that interact with the neural.[26]

For the development of extended realities (XR), in order to blend features of AR and VR, research is being developed by companies with the aim of implementing functionalities that seek to interact directly with all human senses and, on many levels, directly with neural activity.

One of these functionalities is the device's ability to track individuals' gaze and react to the focus of vision, interacting directly with neural activity.[27] Sony launched in 2023 its PlayStation VR2, which already has this type of functionality[28] and reaches a mass target of consumers focused on entertainment.

A *streamer* of games gained the attention of the community of digital games[29] at the end of 2023 and beginning of 2024 by playing several titles using a neurotechnology device attached to her head for control in games, in replacement of keyboard and mouse or other traditional devices, demonstrating the growing interest in new forms of entertainment that combine technology and brain activity.

---

[24] Emotiv Inc (2024). Official website: https://www.emotiv.com (accessed on 07.11.2024).

[25] Neurosky (2024). Official website: https://neurosky.com (accessed on 07.11.2024).

[26] See Capoot (2024).

[27] Greenberg and Jerome (2021), p. 17: Eyes may divulge information about individuals' attention and perhaps even subconscious thoughts in ways that are difficult, if not impossible for individuals to control. Research has shown that pupil dilation, which is relatively easily measured by eye-tracking technologies, indicates an individual's level of interest in what they are looking at and can lead to conclusions—whether accurate or not—about a user's age, gender, or race.101 Gaze can also be used to diagnose medical conditions; eye movements and pupil response can be used to diagnose ADHD, autism, and schizophrenia. Perhaps most concerning, eye tracking technologies can be used to ascertain conclusions about a user's sexual attraction (…) The information divulged from eye tracking can be augmented with other sensors, including movement sensors, EEG and brain-computer interfaces (BCIs), and other pressure and fitness sensors.

[28] Sony PlayStation (2023).

[29] Carpenter (2024).

In recent years, some uses of neurotechnologies for entertainment that were discussed in academic articles have become a reality, such as the use for electronic games, demonstrating that there is a clear interest from companies in exploring the relationship of the human brain with machines as the next generation of technologies.

References

Andorno R, Ienca M (2017) Towards new human rights in the age of neuroscience and neurotechnology. Life Sci Soc Policy 13:5. https://doi.org/10.1186/s40504-017-0050-1

Bear MF, Connors BW, Paradiso MA (2017) Neurociências: desvendando o sistema nervoso, 4ª ed. Artmed editora, Porto Alegre

Bowditch S (2022) Cochlear implant surgery and rehabilitation. Johns Hopkins Medicine. https://www.hopkinsmedicine.org/health/treatment-tests-and-therapies/cochlear-implant-surgery. Accessed 07 Nov 2024

Capoot A (2024) Neuralink rival Synchron's brain implant now lets people control Apple's Vision Pro with their minds. CNBC Tech. https://www.cnbc.com/2024/07/30/neuralink-rival-synchron-offers-thought-control-with-apple-vision-pro-.html. Accessed 07 Nov 2024

Carpenter N (2024) Twitch Streamer plays Palworld using only her mind. Polygon. https://www.polygon.com/24054433/twitch-streamer-palworld-mind-control-eeg. Accessed 07 Nov 2024

Gomez-Gil J, Nicolas-Alonso LF (2012) Brain computer interfaces, a review. Sensors 12:1211–1279. https://doi.org/10.3390/s120201211

Greenberg J, Jerome J (2021) AUGMENTED REALITY + VIRTUAL REALITY: privacy & autonomy considerations in emerging, immersive digital worlds. Future of Privacy Forum. https://fpf.org/wp-content/uploads/2021/04/FPF-ARVR-Report-4.16.21-Digital.pdf

Homer ML et al (2013) Sensors and decoding for intracortical brain computer interfaces. Annu Rev Biomed Eng 15:383–405. https://doi.org/10.1146/annurev-bioeng-071910-124640

Information Commissioner's Office (2023) ICO tech futures: neurotechnology. London, UK, licensed under the Open Government Licence. https://ico.org.uk/media/about-the-ico/research-and-reports/ico-tech-futures-neurotechnology-0-1.pdf. Accessed 07 Nov 2024

Liv N (2021) Neurolaw: brain-computer interfaces. Univ St. Thomas J Law Public Policy (Minnesota) 15:328–335

Lynch Z (2006) Neuropolicy (2005–2035): converging technologies enable neurotechnology. Creating new ethical dilemmas. In: Bainbridge W, Roco M (eds) Managing nano-bio-info-cogno innovations, 1st edn. Springer Dordrecht.

Naysmith C (2023) What do Elon Musk, Bill Gates, Jeff Bezos and retail investors have in common? They're investing millions into brain-tech startups. Yahoo! Finance. https://tinyurl.com/2k39vdvm. Accessed 07 Nov 2024

Sententia W (2004) Neuroethical considerations: cognitive liberty and converging technologies for improving human cognition. Ann N Y Acad Sci 1013:221–228. https://doi.org/10.1196/annals.1305.014

Shenoy P et al (2007) Generalized features for electrocorticographic BCIs. IEEE Trans Biomed Eng 55:273–280

Sony PlayStation (2023) Official website. https://blog.playstation.com/2023/02/06/playstation-vr2-the-ultimate-faq/. Accessed 07 Nov 2024

Steryl D et al (2016) On similarities and differences of invasive and non-invasive electrical brain signals in brain-computer interfacing. J Biomed Sci Eng 9:393–398

The Royal Society (2019) iHuman: blurring lines between mind and machine. https://royalsociety.org/-/media/policy/projects/ihuman/report-neural-interfaces.pdf. Accessed 06 Nov 2024

Chapter 3
Neurorights

To analyze whether the current Brazilian constitutional construction allows for protection against threats arising from neurotechnologies, it is necessary to study the proposals of neurorights as structured normative systems for these new risks. Therefore, this chapter aims to identify the legal object to be protected, through the main theoretical proposals on the subject in the international scenario. From the identification of the object of protection, the new risks that neurotechnologies pose to this legal object will be analyzed. In addition, the arguments about the reach of the main legal structures that currently exist in relation to the object, which would be the object of protection, will be verified. Next, the chapter will address the construction of neurorights that was adopted in Chile and that is based on the systematization of Rafael Yuste, with the respective criticisms. Subsequently, considering the fragility of the legal reasoning pointed out in the criticisms of this model, the model of Marcello Ienca and Rafael Andorno, which is considered a proposal normative of better theoretical structure, will be studied.

3.1 The Legal Object Protected: Neural Data

First, to understand the object that neurorights seek to regulate, it is necessary to establish a concept that can be considered a legal object capable of being protected. Even though it is not exhaustive and completely delimited, it is important to determine a minimum concept, since different types of data attract specific legal protections, for example, the General Data Protection Law in Brazil, which applies to data processing and classifies some special categories of data as sensitive personal data.[1]

[1] II—Sensitive personal data: personal data on racial or ethnic origin, religious belief, political opinion, affiliation to a union or a religious, philosophical, or political organization, data related to health or sexual life, genetic or biometric data, when linked to a natural person.

A. A. Possa, *Neurorights and the Era of Neurotechnologies*, SpringerBriefs in Law, https://doi.org/10.1007/978-3-031-92002-8_3

In Chile, Senate Bulletin 13827-19, which proposed the constitutional insertion of neurorights as a new scope of protection of human dignity, justifies that the object of protection is the content of brain activity, since neurotechnologies go beyond the brain as a physical organ:

> It is for this reason that **the constitutional consecration of the right to neuroprotection** derives from the need to protect human dignity in the face of the use of new techniques, especially in regard to the protection of the "human brain," a concept that is not exhausted only in a physical dimension, but rather expands towards its dimension of mental potentiality that envelops the mysteries of human existence and it is for this reason that it must have the maximum protection ius fundamental (emphasis in the original).[2]

To identify what would be the dimension that goes beyond the physical for the object of protection of neurorights, the current theoretical proposals are based on the human substrate that neurotechnologies end up having as their main focus of interaction: neural data or brain data.

It is important to note that the terms "neural data" and "brain data" are treated as synonyms.

According to Hallinan et al., neural data are those that directly represent the functions of the human brain.[3] In a similar conceptualization, Yuste et al. define brain data as quantitative data about the structure, activity, and functioning of the human brain.[4]

The Information Commissioner's Office in the United Kingdom defines neural data as "first *order data gathered directly from a person's neural systems (inclusive of both the brain and the nervous systems) and second order inferences based directly upon this data.*"[5]

Relating these conceptions to the explanation of the relationship of the central nervous system (brain and spinal cord) with the peripheral systems, cited in Sect. 2.2, we arrive at an understanding that neural data are manifestations of the activities that occur in these systems. This relationship shows that the term studied has a wide variety of uses, both in different contexts of the human body and at different levels (genetic, molecular, etc.), and that it can be decoded in different ways. In this sense, there are several relationships[6] between different neural activities that cannot yet be explained at the level at which neural data are being studied.

Abel Wajnerman Paz comes to a similar conclusion about the concept of neural data, but starting from different premises. For him,[7] neural data are responsible for transfers of information in an inorganic dimension that is related to psychological states, unlike other means that have an organic nature, such as organs and tissues, which are protected by rights related to health.

---

[2] Senate of the Republic of Chile (2020).

[3] Hallinan et al. (2014), p. 55.

[4] Ienca M et al. (2022), p. 20.

[5] Information Commissioner's Office (2023), p. 8.

[6] Frégnac (2017), p. 2.

[7] Wajnerman Paz (2022), p. 401.

The fact is that this author's construction is for the protection of privacy, demonstrating that it is a new dimension that is not protected by already recognized dimensions, and therefore, needs to be linked to the possibility of identifying individuals, ignoring situations where risks may exist without the possibility of individual identification.

The difference in conceptualizations lies in the factor of identification, because for Yuste et al., the neural data that should be the object of neurorights as a broad category are quantitative, that is, they do not depend on qualities such as the possibility of identifying an individual, while for this Chilean author the concept needs to be linked to the possibility of identification.[8] He starts from a philosophical vision of the cognitive sciences of the twenty-first century that cognitive mechanisms are constituted by structural and functional properties that cannot be understood as ontologically independent of the properties of the biological structure.

Neural data, as part of neural information and neural computation, are relatively independent of organic material in the sense that they are not transported through biological materials, but they have a structural dependence as a consequence of the neurocognitive system as a whole.[9]

They can carry non-natural semantic information (*non-natural semantic information—nNSI*), which are information not directly correlated to the content it is dealing with (he cites the example of a broken watch arm, which does not accurately display the time of day, but by non-natural convention can be related to the indication of a time) and natural semantic information (NSI), which are information directly linked to the structure they refer to. In this second case, the information that neural data transfer is always personal, as it relates to characteristics of the individual and personal structure, and that can allow identification.

Thus, similar to Yuste et al., the concept of neural data is that they are "*fundamentally personal information about neural states, processes and structures. They are information about properties within this domain.*"[10]

Again, it is important to highlight that the construction of this concept is made for the regulation of neural data as an object of protection in a new dimension of privacy. Even the author recognizes that neural data can carry non-natural semantic information, because a certain pattern of reading the frequency of neural activity through EEG indicates that there is a brain being observed, because the pattern is the same for all human brains.

However, Yuste et al. propose a taxonomy of neural data that demonstrates the possibility of harm even in cases where they do not allow identification, because they transfer information that, if manipulated, can even lead to physical damage (see Fig. 3.1).

---

[8] Wajnerman Paz (2022), p. 403: The main reason why we focus on personal neural data protection is that the most prominent contexts in which neural data can be used to harm us, mostly by enabling discriminatory or stigmatizing behavior, are contexts in which our information can be linked specifically to us.

[9] See Wajnerman Paz (2022), p. 401.

[10] Wajnerman Paz (2022), p. 402.

The first category (contextual data) refers to methods for direct reading of the electrical activity associated with neural activity, the second (indirect measures of neural structure/function) is associated with readings of neural activities supported by metabolic activities and the last (direct measures of the system/neural function), through digital phenotyping[11] active or passive, generates data about perception, cognition, emotion, and behavior.

This taxonomy demonstrates that neural data have different functions depending on the context in which the body's activity is happening, which can generate a wide variety of consequences. Such consequences range from the identification of an individual, for example, through the decoding of signals that allow the reading of a stimulus,[12] to the enabling of lost human senses.[13]

Thus, as a legal object to be protected through normative systems, it is possible to infer that neural data:

1. They concern not only individual identity, but also perform motor, sensory, emotional, and behavioral tasks.
2. Their content depends on the context of the human body that is related.
3. Neuroscience has not yet been able to map and explore all the activities that neural data perform when correlations are analyzed with different parts of the human body, and normative proposals should consider new possibilities.

The aforementioned conclusions permit the examination of the novel risks that neurotechnologies engender. This is because it is possible to investigate the application of extant legal safeguards through an analysis that incorporates a variety of protected legal goods.

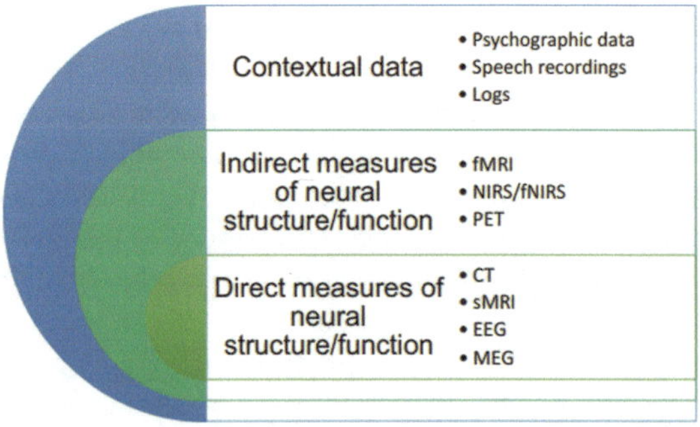

Fig. 3.1 Taxonomy of brain data [Source: Ienca M et al. (2022), p. 20]

[11] See Torous et al. (2016), digital phenotyping is: *moment-by-moment quantification of the individual-level human phenotype in-situ using data from smartphones and other personal digital devices.*

[12] As is the case with lie detection systems based on fMRI technologies described by Langleben and Moriarty in the article "*Using brain imaging for lie detection: Where science, law, and policy collide.*" See Langleben and Moriarty (2013).

[13] Cases of mechanical eyes that were produced by the company *Second Sight Medical Products* and that allowed someone without vision to see again.

3.2 New Risks

The structuring of neurorights theories has its origins in concerns arising from the factual scenario described in the first chapter.

There are two fundamental concerns arising from neurotechnologies related to the definition of the human being itself, from the moment that their body is no longer their limit and what the human-machine entity will be the bearer of human rights.[14] In concrete terms, the main risk is the loss of control of brain functions through the use of technologies or by third parties.[15]

It is possible, from what has been narrated, to delimit the following premises: (1) neurosciences are evolving rapidly, allowing a deeper understanding of brain functioning; (2) neurotechnologies are being developed as the next big technological revolution; and (3) technology companies are seeking to expand the reach of neurodevices to the consumer market with services and products based on brain interactions.

These premises should be analyzed in conjunction with other scenarios related to the current state of technology.

The utilization of big data techniques by private enterprises and public authorities has seen a marked increase in recent times. These techniques are employed to process vast and varied sets of data,[16] facilitating the extraction of valuable insights derived from the information. A notable application of this is the understanding of individual preferences and characteristics, which enables informed decision-making processes. Artificial Intelligence algorithms are reaching levels of independence that allow for automated decision-making and even the programming of new algorithms without human interference, mainly through neural networks.[17]

The evolution of the Internet structure, with the fifth generation of mobile Internet networks (5G), which is in the process of implementation in many countries, will enable the increase in connection density, transmission rates, low latency, the amount of data transmitted by electromagnetic spectrum and energy efficiency, enabling new types of products and services based on interconnected devices (also known as the Internet of Things).

Neurotechnologies used in the context of these technologies will allow, for example, the mass collection of neural data. With the advancement of neuroscience network models and machine learning techniques, it is likely that, in the long term, the analysis of brain data will lead to a greater revelation of mental information. Big data approaches that combine brain data with contextual information can offer additional inferential resources for predictive analyses, allowing more comprehensive and personalized inferences, especially related to mental content. The mental decoding has the potential to enhance our scientific understanding of mental illnesses and

[14] Albers (2018), p. 20.

[15] Albers (2018), p. 20.

[16] Hoffman-Riem (2021), p. 19.

[17] See Hoffman-Riem (2021), p. 17.

offers promises for the targeted modulation of mental states. However, this also raises important questions about privacy and security.[18]

The collection and processing of large neural databases by Artificial Intelligence algorithms for predictions could generate complex inferences about, even, mental content. Another example mentioned above is the possibility of modulating mental states, that is, literally manipulating thoughts and neural processes.

In addition to the uses of brain data itself, these data could be combined with other types of data categories, such as data collected on social networks, genetic data, and geolocation data, making it possible to conduct even more complete analyses about individuals.

The use of neural data in these contexts poses risks to fundamental rights, and the scope of current constructions in the face of this new scenario, whose neurorights are proposed below as a structure for the protection of legal goods related to the activity of neural data, through the use of neurotechnologies.

In a mapping of how academic proposals for neurorights are conditioned to the protection of ethical values, the following taxonomy is presented to systematize (see Fig. 3.2).

Each proposal for neurorights in international academic studies is presented as a reaction to possible damages against ethical values, being them: privacy, freedom of thought, mental integrity, personality, protection against discriminatory biases, and fair access to improvements by neurotechnologies.

Below, the values are systematized according to two large groups of risks: to privacy and to freedom, integrity, and autonomy.

3.2.1 *Privacy*

The collection and access to neural data is a great asset, allowing deep knowledge about the individual sphere at a level that until today has not been possible through other technologies and techniques. A neural profiling[19] deep understanding of people may reveal information unknown even to themselves.

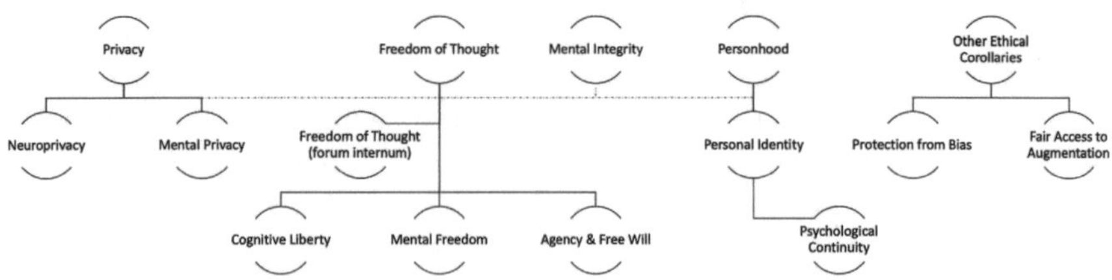

Fig. 3.2 A taxonomy of neurorights [Source: Ienca (2022), p. 6, © 2021 Ienca]

[18] Ienca et al. (2022), p. 3.

[19] Rainey et al. (2020), p. 2303.

A study showed, through the application of neural image reading methodologies (*neuroimaging*), with the purpose of trying to identify individual traits, that there is a real and concrete possibility, albeit limited, of inferring personal data through images of neural activity.[20]

More recently, it was possible to recreate the song *Another Brick in the Wall* by Pink Floyd from *datasets* of iEEG (intracranial electroencephalography) from 29 patients, allowing the generation of information through the analysis of neural activity.[21]

From these examples, it is possible to deduce two major difficulties for mental privacy with the processing of brain data.

The first difficulty concerns the individual's ability to filter[22] the flow of information and segregating private information, a difficulty arising from the fact that brain data do not allow separation between the data that are processed and the system that makes decisions about this treatment (the human brain). At this point, it is possible to verify that neural data has a dual functionality: as personal data that allows the identification of the individual and as a means of communication with other structures of the human body (for example, when the brain sends commands for the peripheral nervous system to perform an action).

The second difficulty concerns the fact that brain information is the last source of individual privacy,[23] since it includes unexecuted actions or unexpressed ideas. Such elements, until now, are not detectable by conventional means, whether in situations of informational surveillance or digital tracking. This lack of individual control can be understood as an impossibility of exercising informational self-determination within the context of neurotechnologies, which, by their nature, prevents the application of observance measures to this safeguard.

However, neurodevices may not be optional, but rather a requirement (such as for military and educational use), or even a social necessity (as currently the use of *smartphones* is a necessity for social relations). From this perspective, individuals will lose total control over extremely intimate elements of their privacy and private life.

Also, there will be problems related to information security that will put neural privacy at risk. One of the possibilities is the attack called *brainhacking*,[24] which allows aggressive and non-consensual access to intimate life, at its most exclusive level, such as thoughts, by criminals.

Some examples are already cited as current concerns. QianQian Li, Ding Ding, and Mauro Conti divide the attacks according to the context of the use of

[20] See Farah et al. (2009).

[21] See Bellier et al. (2023).

[22] Ienca (2021), p. 30.

[23] Ienca (2021), p. 31.

[24] See Haselager and Ienca (2016).

neurotechnologies:[25] medical use; use in electronic games and for entertainment; use for authentication; and on smartphones.

Attacks in the medical use of neurotechnologies include the interruption of the transmission of neural data between a neurodevice and a doctor for the calibration of a prosthetic limb, allowing the attacker to gain access to the transmitted data.

In the context of using neurodevices for user authentication, the authentication system would cause a loss of privacy through attacks using the replication of EEG waves.

The use of neurotechnologies in electronic games and for entertainment has a strong relationship with the context of this work, considering the analysis presented of the developments of companies in the area that seek new services and products based on neural data. The operation of neurodevices for games has APIs to allow access to neural data by the system, generating the possibility of attacks on the APIs and creating a transmission channel for the data to the origin of the attack.

Finally, neurotechnologies in smartphone applications are subject to attacks targeted at the device itself, since the neural data can be stored in its memory, which allows the attacker to extract this data.

These possible risks that neurotechnologies represent may be beyond the reach of current constructions of privacy and personal data protection. This is because, unlike other types of data that only allow identification, neural data can also generate consequences of other orders, considering the correlations that can occur when they are in the context of the functionality of other organs and parts of the human body.

3.2.2 Freedom, Integrity, and Autonomy

Contrary to the capacity of data-encoding technologies to facilitate the interpretation of information, neurotechnologies have the potential to not only decipher the content of data but also exert a direct influence on its functionality. This capacity for direct manipulation can result in the infringement of rights pertaining to individual autonomy, particularly with regard to physical and mental well-being.

This concern about the ability of neurotechnologies to affect aspects of freedom (many of the legal dimensions of which are considered internationally recognized human rights) emerged in some of the earliest theories about neurorights by Wrye Sententia, as early as 2004. According to him, psychiatric therapies[26] that modify mood based on transcranial magnetic stimuli needed a study on the possibilities and limitations of uses, since these therapies can be considered mandatory in specific cases and reflect on an individual's freedom.

[25] See Conti et al. (2015), pp. 664–665.

[26] Sententia (2004), p. 224.

The field of neuromarketing, which has existed for several decades, originated from research that identified brain reactions to some different stimuli. Ienca and Andorno cite the emergence of this area as a branch of neuroscience and as a consequence of some experiments, such as one in which individuals were subjected to fMRI techniques while drinking Coca-Cola consciously and individuals consuming the product unknowingly, to demonstrate that there were differences in the brain response of consumers.[27] Companies such as Google, Disney, and CBS began using research with *neuromarketing* to understand the unconscious preferences of consumers and incorporate marketing strategies that could interact with them.

It is important, at this point, to differentiate the effects of neurotechnologies from what is known as *nudges*, considering that both have in common the possibility of manipulating human behavior. According to Thaler and Sustein, *nudges* are "any *aspect of the choice architecture that alters people's behavior in a predictable way without forbidding any option or significantly changing their economic incentive*"[28] which are employed in the digital world to induce consumer behaviors through different mechanisms.

Through research in 71 scientific articles, Ana Caraban and Evangelos Karapanos identified 23 mechanisms used to influence the decision to consume with technologies.[29] Such mechanisms were classified into four categories: facilitate, confront, deceive, and social influence. The mechanisms of *nudge* are based on the idea that the individual can be influenced to make decisions based on the possibility of exploring how decision-making happens.

Neurotechnologies, on the other hand, would be more effective than *nudges* because they have the power to directly interfere with brain activity through interaction with neural data through equipment specifically created for this purpose.

As an example of the level at which these technologies can alter mental integrity, it is possible to cite research that involved BCIs aimed at manipulating emotions to modify an individual's emotional state from neutral to happy, or, then, to alleviate painful memories of traumatic events in a kind of "emotional prosthesis."[30]

For entertainment, some BCIs capture emotions during the reading of a book[31] and then, based on the change in the reader's emotional state, automatically adapt the version of the content that is shown to the individual.

In these cases, there would not only be *nudges*, but also the direct influence on the mental autonomy of individuals, removing the individual freedom to decide about their own emotions. For example, Steinert and Friedrich[32] reinforce the concern about BCIs with these capabilities in situations like the scandal of *Cambridge*

---

[27] Andorno and Ienca (2017), p. 4.

[28] Sunstein and Thaler (2008), p. 6.

[29] See Caraban and Karapanos (2020).

[30] Friedrich and Steinert (2020), p. 353.

[31] Friedrich and Steinert (2020), p. 354.

[32] Friedrich and Steinert (2020), p. 360.

Analytica in the United States presidential elections in 2016. The company used databases from the social network Facebook to manipulate political opinions.

Another example is the neurotechnologies used for the improvement of soldiers in the military area. The military uses of brain-computer interfaces (BCIs) include the collection of neural information from the brains of fighters to adapt and improve their equipment in a personalized way, in addition to the development of threat alert systems that transform subconscious neurological responses to danger into consciously accessible information.[33]

Just as new data processing technologies have raised concerns, by taking the level of information inference to a new level, neurotechnologies have the potential to allow external influence on human decision-making at a level that can lead to total loss of mental self-determination. This level may be beyond the protection conferred by the current scope of the right to psychic integrity.

Furthermore, within the domain of potential violations of freedom and individual autonomy, we observe the emergence of BCIs that can challenge the boundaries of the right to physical integrity. These boundaries pertain to both the level of brain structure and the realm of regulating the actions of the human body in its entirety.

As highlighted earlier, neural data can be responsible for performing motor tasks if they interact with the peripheral nervous system. Currently, some of the most famous BCIs are those related to the recovery of movements by individuals with severe motor limitations, such as the exoskeleton used by a paraplegic young man to kick the ball in a soccer game at the 2014 World Cup in Brazil.

Another example concerns BCIs that allow the recovery of senses by people who have suffered irreversible damage to the organs responsible for them. In this sense, the mechanical eyes of the company *Second Sight Medical Products* have allowed people without vision to see again.

The risks here can be of various orders, such as malicious actions by third parties or even the discontinuation of the service, which removes the vision capacity of Second Sight device users.

Risks related to malicious actions are explored by Ienca and Haselager as neurocrimes.[34] Such crimes refer to actions that seek to modify or interrupt the operation of equipment that interfaces with the brain, being able to interact directly or indirectly with brain data, without necessarily having access to neural data by criminals.

As a type of neurocrime, the authors elaborate on the crimes of *brain-hacking*, which are composed of three characteristics: *(1) can only be performed on neural devices that establish a direct connection pathway with the brain such as tDCS, neural implants and BCI, (2) involve the direct access to and manipulation of neural information, (3) influence directly neural computation in the users.*[35]

These crimes, which occur through actions that exploit vulnerabilities in BCIs, seek to gain access to neural data with the possibility of causing total loss of

---

[33] Rickli and Ienca (2021), p. 209.

[34] See Haselager and Ienca (2016).

[35] See Haselager and Ienca (2016).

physical autonomy to users, since the criminal could perform physical actions using the victim's limbs to, for example, physically attack third parties.

In the previously mentioned hypothesis of BCIs that reactivate senses that have been debilitated, a case from the company responsible for the development and maintenance of the mechanical eyes details some risks, such as the loss of senses because of a business decision that opts for the discontinuation of the product, or the bankruptcy of the institution itself. In 2019, the implant was discontinued, and in 2020, the company almost went extinct due to financial problems, and in 2022, it was the subject of a merger with another company, which, in turn, did not commit to the continuity of updates and maintenance of these devices.

In this case, users are at risk of losing their vision due to financial and corporate decisions, given that defects may arise in the devices and there are no other institutions with knowledge for proper maintenance.

Both scenarios demonstrate risks that take offenses to the right to physical integrity to unprecedented situations, such as the possibility of completely losing control of your body, which can be used to commit criminal acts by a third party, or the loss of senses caused by corporate decisions.

Finally, it should be noted that in addition to the risks generated by some of the situations described above for both rights mentioned in this chapter, there is still the possibility of *brain-hacking*, which not only allows the loss of movements but also the invasion of privacy and data protection.

3.3 The Proposals for Neurorights

First, it is necessary to establish a history of the works that deal with neurorights and are considered as their main study milestones and proponents.

In 2004, Wrye Sententia, in the article "*Neuroethical considerations: cognitive liberty and converging Technologies for improving human cognition*," structured the concept of cognitive freedom as the set of two principles for the protection of threats to mind control by neurotechnologies: (1) The first principle deals with the protection of an individual against being forced to use neurotechnologies or psychoactive drugs; (2) The second principle protects the freedom to use neurotechnologies or drugs that enhance mental capacity, as long as the use is not associated with risks to third parties.[36]

In 2013, Jan-Cristoph Bublitz, in the article "*My Mind is Mine!? Cognitive Liberty as a Legal Concept*," deepened the precept of cognitive freedom as a legal concept. In his words: "*It is, or rather, should be the central legal principle guiding the regulation of neurotechnologies, guaranteeing the right to alter one's mental states with the help of neurotools as well as to refuse to do so.*"[37]

---

[36] See Sententia (2004).

[37] See Bublitz (2022), p. 234.

In 2017, Rafael Yuste, Sara Georing, and colleagues published in the journal *Nature* the article "*Four Ethical Priorities for Neurotechnologies and AI,*"[38] in which they list four areas of ethical concerns to which engineers, technology developers, and academic researchers should be exposed. The main points of this latest proposal will be briefly analyzed.

The first, privacy and consent, concerns the possibility of Internet-connected neurotechnologies being adopted for the use of predictive algorithms by companies, criminals, or governments, for tracking and manipulation purposes. The second, agency and identity, discusses the possibility of technology promoting the breakdown of an individual's notion of identity and agency, modifying core moral conceptions of personal personality. The third, enhancement, or *augmentation,* refers to the possibility of different access to neurotechnologies that enhance brain activity, which would create groups of people with greater brain capabilities compared to non-users of these technologies, resulting in greater inequalities. Finally, the last one is about biases, or *bias,* which deal with discrimination that algorithms can bring based on access to neural data.

These ethical concerns influenced, in Chile, the approval of two proposals from the same group of senators, in 2021: Bulletin No. 13.827-19, which inserts protections for "mental integrity and identity in relation to the advancement of neurotechnologies," and Bulletin No. 13.828-19, which creates legislation on "protection of neurorights and mental integrity and the development of research in neurotechnologies."

The projects aim to address a broad scope of protections against the risks that neurotechnologies pose at the constitutional and infraconstitutional levels.

In the justification for the constitutional insertion, the proponents cite as the content of neurorights the protection of the four areas of ethical concern that were structured by Sara Goering and Rafael Yuste in the journal *Nature*: privacy and consent, self-determination, equality of development, and combating algorithm biases.

However, even though the Chilean senators themselves have pointed out that the minimum content of neurorights are the above-mentioned ethical concerns, they recognize that this content was not exhausted in them. Therefore, when dealing with "open texture law," it should be regulated through the infraconstitutional legislator, with the constitutional text only being necessary to open new prerogatives for the fulfillment of this right as neurotechnologies are developed.

The final wording, after debates with the scientific and political community, was:

> Scientific and technological development will be at the service of people and will be carried out with respect for life and physical and psychic integrity. The law will regulate the requirements, conditions, and restrictions for its use in people, especially safeguarding brain activity, as well as the information derived from it.[39]

This constitutional innovation in Chile attracted criticism from some of the main theorists on the risks of neurorights, notably Jan-Christoph Bublitz.

---

[38] See Yuste et al. (2017).

[39] República de Chile (2021).

In the article "Novel *Neurorights: From Nonsense to Substance*," Bublitz criticizes the current construction of Chile, which was based on the initiatives of the *Neurorights Initiative* through the article published in the journal *Nature* by Rafael Yuste, Sara Goering, and colleagues. The initiative presents five new rights to be recognized as human rights in the international community. According to the critic, the adoption of the theory of five new human rights would mean an inflation of these rights, which could lead to a decrease in their importance.

Among the criticisms listed are:[40] (1) that there is a lack of solidity in the analyses, when it is said that current rights do not have efficacy against challenges from neurotechnologies, (2) that the proposals are developed by researchers in the medical sciences, who have knowledge about the object to be regulated, but not about theories of regulation, and can be considered more the result of activism than of legal analyses, (3) that human rights limit democracy through the creation of duties for the state and, finally, (4) that the object of the proposals for neurorights is not even clear, which, in some cases, is very evident that it is not necessarily related to brain activity, but to other social and behavioral elements. Bublitz states that there is a need to treat all matters related to the brain as demanding new and exceptional protections.

However, he points out that these constructions are based on texts that address ethical concerns. Even though these are legitimate concerns, in terms of rights construction, it is not possible to identify a scientific article with the proposal of these rights that has undergone peer review in the field of human rights or constitutional law. The criticism regarding the lack of normative rigor in the proposals is reinforced by other scholars.[41]

The only construction that seems to be endowed with theoretical rigor, according to Bublitz,[42] and which is also the subject of analysis in other works,[43] is that of Ienca and Andorno.[44]

In 2023, these authors, along with Bublitz, Abel Wejnerman Paz, Sara Goering, and other authors already mentioned, published an article with the aim of unifying understandings about neurorights proposals based on contours of legal theories, and not purely ethical ones, being them: mental integrity, mental privacy, and cognitive freedom.[45]

---

[40] See Bublitz (2022).

[41] See Borbón and Borbón (2021).

[42] See Bublitz (2022), footnote 11.

[43] Farinella and Gulyaeva (2022), p. 284.

[44] Andorno and Ienca (2017).

[45] See Ligthart et al. (2023), p. 463: In this paper, we will not take a position about the need for specific neurorights to protect the core subject matter identified in Section "Conceptual and ethical foundations of mental integrity, mental privacy, and cognitive liberty." Rather, we aim to advance the debate by considering the extent to which mental integrity, mental privacy, and cognitive liberty enjoy protection under the established framework of human rights law (Section "Legal foundations of mental integrity, mental privacy, and cognitive liberty"). This systematic analysis based on the normative foundations discussed in the section "Conceptual and ethical foundations of mental integrity, mental privacy, and cognitive liberty" will be relevant to all three central positions regarding the development of specific neurorights on the international and regional levels.

3.3.1 Cognitive Freedom

The proposal for cognitive freedom was first presented in 2004, identifying its fundamental right contour as "*the right and freedom to control one's own consciousness and electrochemical thought processes is the necessary substrate for just about every other freedom.*"[46]

If we consider that the legal object to be protected is the neural data according to the elements described in Sect. 2.1, we can identify the compatibility of this concept with neural data, in the context of the electrochemical processes responsible for consciousness, which occur in the central nervous system.

According to Bublitz's construction, which tries to identify normative characteristics in cognitive freedom, it would be a dimension that would expand the general freedom of self-determination (*free will),* responsible for protecting the mental sphere against third-party interventions, something that this already recognized general right does not reach, as it is focused on a right of individual development and not on a prohibition of direct interventions in mental processes. In his words, "*Cognitive liberty is, in a way, the right to free will, protecting the conditions of possibilities of 'free' actions and therewith of blame and retribution.*"[47]

One of the premises of this author for the elaboration of a legal concept of cognitive freedom concerns the insufficiency in the reach of legal concepts such as "harm" for the particularities of the brain/mind structure, considering that the legal idea of harm is always linked to bodies, demanding that legal protections be created for the mind that are not identical to those existing for the human body. He summarizes: *The point is this: the lack of a theoretical framework of negative interventions into other minds is entwined with the lack of considerations on positive, self-determined alterations of one's own mind. Normatively, both are two sides of the same coin: Cognitive liberty.*[48]

As legal manifestations that would approach the conception of cognitive freedom, he cites: freedom of conscience (*freedom of thought*), personality and privacy rights, and the right to mental integrity.

However, for this author, the rights in question are not adequate to protect the particularities of neurotechnologies based on the premise that their elaborations never needed to take into account criteria for direct interference in mental activity, lacking a legal principle that guarantees the possibility of more concrete regulations to protect the mental sphere against third-party interventions.

The dimensions of this right are: (1) the freedom to change one's own consciousness, allowing the retention or elimination of any mental state and the exercise of mental capacities and (2) protection against the imposition of other wills by third parties, whether through coercive use or prohibition of use of neurotechnologies. Thus, two obligations arise: a negative one, of non-interference in mental processes

---

[46] Andorno and Ienca (2017), p. 10.

[47] Bublitz (2022), p. 242.

[48] Bublitz (2022), p. 243.

through the manipulation of neural data, and a positive one, which obliges states to promote cognitive freedom.

In terms of effectiveness, it is not an absolute right,[49] being possible interference in the face of conflict of other rights or interests, but that the analysis depends on the concrete case through criteria such as: the consequences of effects (positive or negative) and the intensity of interferences.

For Ienca and Andorno, as neurocognition is the substrate of all the other freedoms, as it accompanies the mental processes underlying human consciousness, it cannot be reduced to existing rights.[50]

Thus, cognitive freedom would be an effective protection against any and all imposition of external will on the electrochemical processes responsible for neural data related to thoughts, mood states, and behaviors, through the forced use or prohibition of the use of neurotechnologies.

3.3.2 *Mental Privacy or Neuroprivacy*

The techniques mentioned in Sect. 2.2, which address the risks that neurotechnologies engender, encompass novel prospects for privacy violations through technologies that possess the capability to decipher thoughts and intentions by merely reading neural data manifested through magnetic stimuli.

For Ienca and Andorno,[51] a right to mental privacy would arise segregated from the right to privacy and protection of personal data, considering that neural data concerns the most intimate sphere of private life and personality, as well as the way they are collected. The big gap with a general right to privacy and data protection would be the fact that neurotechnologies not only have access to data as they manifest, but also to the origin of these data: neural activity.

This right, therefore, has as its legal object of protection the neural data manifested through brain activities that can be read by neurotechnologies.

As a scope of protection, it protects not only manifested thoughts, but also the moment of their conception in the brain structure, even if they are not realized. It also protects manifestations through electromagnetic readings, which can be stored in environments external to the central nervous system (for example, in databases collected by BCIs).

The question of effectiveness, whether absolute or relative, is left open, considering that the breach of neural privacy for access to memories and thoughts can compromise other fundamental rights, such as the right to the presumption of innocence, in addition to that their relativizations could generate states of authoritarian surveillance that would surpass the fiction of *Big Brother* by Orwell. Relativizations of the

---

[49] Bublitz (2022), p. 255.

[50] Andorno and Ienca (2017), p. 10.

[51] Andorno and Ienca (2017), pp. 11–12.

possibility of using neurotechnologies that perform thought readings would constitute an affront to human dignity itself.[52]

Bublitz recognizes the importance of protecting mental privacy, but does not see sufficient arguments that justify the creation of a new manifestation of the right to privacy or the right to protection of personal data,[53] citing, as an example, the protection of sensitive, genetic, and medical personal data, provided for in current personal data protection legislation.

3.3.3 *Mental Integrity*

The right to mental integrity seeks protection in cases where interference with neural data results not only in the loss of privacy, but also when it directly impacts mental computation.[54]

Given the above conceptualization, to identify the specific scope of protection of this right, it is necessary to understand what the content of a mental computation is. Winfried Denk, Kevin L. Briggman, and Moritz Helmstaedter establish a scenario of knowledge about the brain:

> We now possess a lot of information about the brain's molecular and cellular building blocks. We know how neurons use action potentials to transmit information over long distances, how they communicate with each other through synapses and which nonlinear chemical or electrical processes they use when performing computations. Exquisite methods exist to record electrical activity and calcium dynamics in individual cells 1–3 and in small populations of cells 4–6. Powerful modelling tools allow the simulation of cellular and network behaviour, given a set of cellular parameters and the network connectivity. However, although we know what most neuronal cell types 'look' like and, to a lesser extent, what their biophysical properties are, information about circuit structure is lacking in most cases.[55]

According to the panorama described above, currently, the sciences related to the study of the brain at the physical, molecular, and cellular level, already have a considerable degree of information to understand their dispositions and structuring. The problem lies in knowledge about the system as a whole, since the amount of data needed to understand the long-range relationship of these activities may never be able to be generated.

This problem of long-range relationships between the molecular, cellular, and structural elements of the nervous system (central and peripheral) is important, as they are the ones that, when manifested, are part of the process of cognition of the world by the human being.

[52] Farinella and Gulyaeva (2022), p. 290.

[53] Bublitz (2022), p. 10.

[54] Andorno and Ienca (2017), p. 17.

[55] Denk et al. (2012), p. 352.

The question of how man perceives and interprets the world has been the subject of study for hundreds of years, going through philosophical, physiological, and biological science methods. Churchland and Sejnowski synthesize the question in how to understand "*the nature of the 'mirroring' of the outer world by the inner world, and roles of reason and inference in the generation of internal models of reality*."[56]

We are talking about the relationship between structure (the brain), processes (the connections of cellular and molecular activities in a complex system), and manifestations (thoughts, behaviors), which, in the end, are part of an individual's self-perception.

Various currents on the connection between these relationships have emerged over the decades. Churchland and Sejnowski cite: (1) the materialist, which argues that mental states are necessarily reflections of the physical brain state; (2) the behavioral, which identifies in human behavior the center for understanding the mental state not linked to the physical structures of the brain and (3) the identity, which is close to the materialist in defending that mental states are identical to brain states and that neuroscience is responsible for identifying exactly which behaviors correspond to brain functioning.

These currents have led to theories about what would be, effectively, the necessary methods for a construction on the model of human knowledge. They are:[57] (1) the dualist theory, which sees no relationship between neuroscience and psychology, arguing that both areas should be autonomous and that neuroscience will never be able to provide answers for these relationships, and (2) the reductionist theory, which advocates an integration between psychological and neurobiological methods to seek the answers.

A third theory, called the "*connectionist model*,"[58] arises to argue that these relationships can be identified through a model that studies the tasks performed through an architecture composed of units, layers, and columns, being a true system of interconnected networks that refer to the functioning of digital computers.

Considering this theory, mental computation, as a legal object to be protected, would be the manifestation of neural data that relate the units and layers of mental activity in a system, which ranges from physical and biological level activities to behavioral level manifestations (thoughts, actions, emotions etc.).

Thus, the right to mental integrity protects the neural data responsible for mental computation (at any of its stages or levels) against unwanted external interventions. It is a reconceptualization of the right to psychological integrity,[59] which has as a limitation a protection to the right to psychological health, formulated to guarantee normative protection against unwanted interventions that modify mental computation and cause harm to individuals.

---

[56] Churchland and Sejnowski (1990), p. 344.

[57] Churchland and Sejnowski (1990), p. 349.

[58] Churchland and Sejnowski (1990), p. 357.

[59] Andorno and Ienca (2017), pp. 18–19.

For something to be considered a threat to mental integrity, there must be three requirements: (1) direct access and manipulation of neural data; (2) absence of authorization from the technology user; and (3) physical and/or psychological damage as a result.

The difference regarding the scope of protection of the right to cognitive freedom is that, while the latter seeks normative protections for the free development of neural data activities by the individual himself through the voluntary use or non-use of neurotechnologies, mental integrity focuses on protection against damage to the physical and behavioral structure caused by malicious actions of third parties.

The actions that would be subject to this protection would be the neurocrimes of *brain-hacking* and memory engineering techniques, which allow the manipulation of human memories.

3.3.4 Psychological Continuity

The right to psychological continuity has similarities with the right mentioned earlier, since it also seeks to guarantee protections against non-consensual manipulation of neural data in the process of brain computation. In the 2023 article that tries to build a consolidation of neurorights proposals by the main authors in the field, this proposal is not addressed. However, considering the origin of this work in the article that proposes it, it will be maintained for historical purposes.

The difference, according to Andorno and Ienca (2017), lies in the requirement of mental damage.[60]

Psychological continuity seeks to protect the individual's personality from unwanted changes through external manipulations, in a special dimension of the right to identity.[61]

In considering the psychological context under scrutiny, the legal entity to be safeguarded by this right is the personality from a psychological standpoint (the protection of personality as a normative principle is explored below).

Psychology does not possess a singular concept of personality. For instance, one may cite Freud's theory and Jeung's theory as two of the renowned theoretical debates within the domain of psychological science. To further complicate the issue, Calvin Hall, Gardner Lindzey, and John B. Campbell identify five sources of influence: clinical observation, Gestalt, experimental psychology, psychometric tradition, and finally, genetics and physiology.[62] Therefore, conceptualization depends on the theoretical context that is adopted.

---

[60] Andorno and Ienca (2017), p. 20.

[61] Farinella and Gulyaeva (2022), p. 293.

[62] They cite 18 works as the main ones among the divisions presented above, demonstrating the wide diffusion of perspectives and theoretical constructions. See Hall et al. (2000).

Even within this difficulty, when the authors propose to analyze different types of personality theories, they identify requirements that theories necessarily need to observe, being a "set of assumptions regarding human behavior, along with rules for relating these assumptions and definitions to allow their interaction with empirical or observable events."[63]

These elements help in defining personality as a legal object protected in the right to mental integrity, since it is possible to identify in it a human behavior that encompasses a set of assumptions that can be manifested in empirical and observable events, being the result of electrochemical interactions of neural data in the behavioral context.

Thus, psychological continuity aims to protect an individual's behavior against external threats that use neurotechnologies to influence the neural data that make up this behavior possible to be empirically manifested.

As an example of neurotechnology that would threaten psychological continuity, Ienca and Andorno cite an experiment that used transcranial magnetic stimuli to modulate neural data in regions responsible for social prejudices and political and religious beliefs.[64]

It is not defined whether this is an absolute or relative right, since situations such as the use of neurotechnologies to psychologically modify individuals with tendencies for crimes such as pedophilia, serial killers, and rapists could be socially justifiable.

3.3.5 Are Neurorights New Human or Fundamental Rights?

The 2017 Nature magazine text, which lists ethical risks raised by neurotechnologies, does not mention a normative system, even though it is used as one of the main texts for the construction of Chile.

The Neurorights Foundation cites this same construction in a 2022 analysis, prepared by Jared Genser, Stephanie Hermann, and Rafel Yuste, on international instruments that deal with human rights. This construction is attributed to "prominent neuroscientists and neuroethicists."[65]

The intention, according to them, is to demonstrate that there is no consensus on the gaps in the international human rights scenario that justify the creation of new categories and the attempt to systematize these gaps, existing in international treaties, against possible threats of neurotechnologies. They conclude that the international treaties analyzed are unprepared for this new reality and that the United Nations urgently needs to look at the topic.[66]

---

[63] Hall et al. (2000), p. 28.

[64] Andorno and Ienca (2017), p. 21.

[65] Genser et al. (2022), p. 16.

[66] Genser et al. (2022), p. 29.

The report does not elaborate on what the legal objects would be protected by the new human rights and the relationship with the legal objects protected by international treaties. Here, the criticisms of Bublitz in 2022 highlighted earlier are valid, since the concepts are vague and do not effectively demonstrate the existence of gaps that justify the non-application of a certain protection against this new legal object.

In an analysis of the set of these proposals based on ethical premises, it seems that there really is a failure of legal methodology for such human rights. Also, it is possible to verify that the authors do not even deal with what would effectively be constituted as a "human right."

It seems, indeed, that the intention is to include in international charters the five ethical expressions presented in the 2017 Nature magazine article which are indeterminate concepts and generic scope of application to be reported as human rights.

The Information Commissioner's Office report also addresses some conceptual problems about this systematization. Mental privacy, in the way it is proposed by the Neurorights Foundation, requires an absolute right to data deletion, which can create conflicts with other legislations. Already the notion of mental integrity (or psychological continuity) raises questions about the interpretative extent of the term and even the construction of theories about responsibility.

The construction of Ienca and Andorno of 2017, although proposed under the argument of new human rights, has a better foundation, since, as demonstrated, their justification is carried out through the scope of the effectiveness of specific human rights, being able to identify distinct legal objects for new protective tutelages.

Also, they recognize that the accelerated development of neurotechnologies may not necessarily create human rights, it being possible that there are restructurings of traditional rights to specifically address the challenges of new risks.

In this sense, the analysis carried out in each proposal of neuro-right in the subtitles above really seems to demonstrate that, before the object legally termed as neural data, some protection gaps regarding the legal objects protected by existing rights are observed.

However, for the theoretical support of the normative system called neurorights, an analysis through a positive and consolidated constitutional structure can assist in identifying the relevance of this systematization, under penalty of incurring the same problem pointed out by Bublitz (2022): the topic can be considered activism and not legal regulation.

This is because it is before a constitutional structure that it becomes possible to verify interpretive limits, principiological tensions, and legal conflicts, justifying the legal reasoning for new protective constructions. For comparative purposes, Danilo Doneda, when analyzing the various contours of the right to privacy in different legal systems, identifies that several social factors end up impacting the legal content: "When we glimpse this formation process, we highlight its malleability according to various conditioning factors, always the result of the configuration of the values of a certain society."[67]

---

[67] Doneda (2019), p. 108.

Furthermore, even though numerous constitutionalized fundamental rights have their origins in the international recognition of human rights, it was in the factual construction that principles and values gave light to the content later recognized in the international scenario.[68]

3.3.6 The Trial of the First Lawsuit on Neurotechnologies: The Emotiv Case in Chile

The Supreme Court of Chile, in 2023, decided the first cause involving neurotechnologies in a lawsuit against Emotiv. The marketing of the Insight device in the country was questioned, under the argument that the company violates the constitutional provision on neurorights and Law N° 19.628, which also deals with the topic.

The following risks were pointed out: (1) the possibility of re-identification of anonymized data; (2) the possibility of brain-hacking; (3) the unauthorized reuse of neural data; (4) the commercialization of neural data; (5) digital surveillance; and (6) the collection of neural data for non-consented purposes.

The company, in turn, defended itself with the argument that its device is a non-invasive neurotechnology, without therapeutic purpose and used for personal knowledge (autocuantificacion), not being characterized as a medical device.

The Supreme Court highlighted that it is the role of law to anticipate and prevent human integrity in its entirety in the face of advances and possible negative effects of new technologies and, that, for the concrete case, even though the device is not used for medical purposes, it should be evaluated by the health and customs authority (since it was imported without proper inspection) in order to be evaluated if it is within the national normative parameters, mainly in the face of the alleged risks.

It can be inferred from the sentence that the legal basis was the concern with physical and psychic integrity from a unitary understanding of human personality, including the right to privacy. This basis contrasts with the arguments that several new rights would be necessary to assess the risks of these devices.

References

Albers M (2018) Biotechnologies and human dignity. Direito Público 15(82):9–49

Andorno R, Ienca M (2017) Towards new human rights in the age of neuroscience and neurotechnology. Life Sci Soc Policy 13:5. https://doi.org/10.1186/s40504-017-0050-1

Bellier L, Llorens A, Marciano D, Gunduz A, Schalk G et al (2023) Music can be reconstructed from human auditory cortex activity using nonlinear decoding models. PLoS Biol 21(8):e3002176. https://doi.org/10.1371/journal.pbio.3002176

[68] Sarlet (2018) mentions, for example, that it was in the Magna Carta of England in 1215 that the conception of classic civil rights such as *habeas corpus,* due process of law, and the property was guarantee.

Borbón D, Borbón L (2021) A critical perspective on NeuroRights: comments regarding ethics and law. Front Human Neurosci 15:703121

Bublitz JC (2022) Novel neurorights: from nonsense to substance. Neuroethics 15(1):7

Caraban A, Karapanos E (2020) The 23 ways to nudge framework: designing technologies that influence behavior subtly. Interactions 27(5):54–58

Churchland PS, Sejnowski TJ (1990) Neural representation and neural computation. Philos Perspect 4:343–382

Conti M, Ding D, Li QQ (2015) Brain-computer interface applications: security and privacy challenges. In: 2015 IEEE conference on communications and network security (CNS). IEEE, 2015, pp 663–666. https://doi.org/10.1109/CNS.2015.7346884

Denk W, Briggman KL, Helmstaedter M (2012) Structural neurobiology: missing link to a mechanistic understanding of neural computation. Nat Rev Neurosci 13(5):351–358

Doneda D (2019) Da Privacidade À Proteção De Dados Pessoais, 3ª Ed. Thompson Reuters, São Paulo

Farah MJ et al (2009) Brain imaging and brain privacy: a realistic concern? J Cogn Neurosci 21(1):119–127. https://doi.org/10.1162/jocn.2009.21010

Farinella F, Gulyaeva EE (2022) Human neuro-rights. Revista Quaestio Iuris 15(1):278–299

Frégnac Y (2017) Big data and the industrialization of neuroscience: a safe roadmap for understanding the brain? Science 358(6362):470–477. https://doi.org/10.1126/science.aan8866

Friedrich O, Steinert S (2020) Wired emotions: ethical issues of affective brain–computer interfaces. Sci Eng Ethics 26(1):351–367

Genser J, Herrmann S, Yuste R (2022) International human rights protection gaps in the age of neurotechnology. NeuroRights Foundation. https://static1.squarespace.com/static/60e5c0c4c4f37276f4d458cf/t/6275130256dd5e2e11d4bd1b/1651839747023/Neurorights+Foundation+PUBLIC+Analysis+5.6.22.pdf. Accessed 11 Nov 2024

Hall CS, Lindzey G, Campbell JB (2000) Teorias da personalidade. Artmed Editora

Hallinan D et al (2014) Neurodata and neuroprivacy: data protection outdated? Surveill Soc 12:55–72

Haselager P, Ienca M (2016) Hacking the brain: brain–computer interfacing technology and the ethics of neurosecurity. Ethics Inf Technol 18(2):117–129. https://doi.org/10.1007/s10676-016-9398-9

Hoffman-Riem W (2021) Teoria geral do direito digital: transformação digital: desafios para o direito. Forense, Rio de Janeiro

Ienca M (2021) Common human rights challenges raised by different applications of neurotechnologies in the biomedical field. Report commissioned by the Council of Europe

Ienca M et al (2022) On neurorights. Front Hum Neurosci 15:701258. https://doi.org/10.3389/fnhum.2021.701258

Information Commissioner's Office (2023) ICO tech futures: neurotechnology. London. https://ico.org.uk/media/about-the-ico/research-and-reports/ico-tech-futures-neurotechnology-0-1.pdf. Accessed 07 Nov 2024

Langleben DD, Moriarty JC (2013) Using brain imaging for lie detection: where science, law and research policy collide. Psychol Public Policy Law 19(2):222–234. https://doi.org/10.1037/a0028841

Ligthart S, Ienca M, Meynen G et al (2023) Minding rights: mapping ethical and legal foundations of 'neurorights'. Camb Q Healthc Ethics 32(4):461–481. https://doi.org/10.1017/S0963180123000245

Rainey S et al (2020) Brain recording, mind-reading, and neurotechnology: ethical issues from consumer devices to brain-based speech decoding. Sci Eng Ethics 26(4):2295–2311. https://doi.org/10.1007/s11948-020-00218-0

República de Chile (2021) Ley 21383. https://www.bcn.cl/leychile/navegar?idNorma=1166983. Accessed 21 Nov 11 2024

Rickli JM, Ienca M (2021) The security and military implications of neurotechnology and artificial intelligence. Clinical neurotechnology meets artificial intelligence: philosophical, ethical, legal and social implications, pp 197–214

Senado de la República de Chile (2020) Boletín 13827-19. Santiago, Chile

Sententia W (2004) Neuroethical considerations: cognitive liberty and converging technologies for improving human cognition. Ann N Y Acad Sci 1013:221–228. https://doi.org/10.1196/annals.1305.014

Sunstein CR, Thaler RH (2008) Nudge: improving decisions about health, wealth, and happiness. Yale University Press, London

Torous J et al (2016) New tools for new research in psychiatry: a scalable and customizable platform to empower data driven smartphone research. JMIR Mental Health 3(2). https://doi.org/10.2196/mental.5165

Wajnerman Paz A (2022) Is your neural data part of your mind? Exploring the conceptual basis of mental privacy. Minds Mach 32(2):395–415. https://doi.org/10.1007/s11023-021-09574-7

Chapter 4
The Protection of Human Dignity and the Scope of Protection of Neurorights

The chapter will initially make brief considerations about the theory of digital constitutionalism and its background, which motivates this international academic effort, since discussions about neurorights can be considered new manifestations of this type of constitutional approach. Then, the development of the protection of human dignity in Brazil is studied, especially its realization in the form of fundamental rights. For this, rights already established in the Brazilian constitutional jurisdiction, as a result of the need to protect dignity, will be analyzed. Finally, the relationship between elements of theories of neurorights and fundamental rights before the Constitution of Brazil will be verified.

4.1 The Role of Law in Protecting Against the Risks and Damages Caused by New Technologies

As already mentioned, it is the values of a society that shape the content of a certain legal construction, justifying or not the structuring of innovations.

This was the reasoning used in the conception of the fundamental right to informational self-determination by the German Constitutional Court, in the case of the 1982 census (BVerfGE 65), when it was confronted with new risks to privacy arising from recent models of personal data processing. The Court found, after identifying gaps in its privacy protection system, that:

> from art. 2, §1 (free development of personality), combined with art. 1, §1 (human dignity), of the LF, would result in a fundamental right to informational self-determination that would guarantee 'the power of the individual to determine the collection and use of his personal data'. (BVerfGE 65,1 (43), Census).[1]

The German Constitutional Court also adopted the same line in relation to the construction of the fundamental right to confidentiality and integrity of personal information systems,[2] when analyzing the judgment of 27.02.2008—1 BvR 370/07, 1 BvR 595/07. Based on the scope of protection of the right to informational

[1] Mendes (2020), p. 10.

[2] Hoffmann-Riem (2021), p. 475.

A. A. Possa, *Neurorights and the Era of Neurotechnologies*, SpringerBriefs in Law, https://doi.org/10.1007/978-3-031-92002-8_4

self-determination, at the time of the judgment, and of the then recognized fundamental rights, given the types of risks that online search engines brought to society, a gap was identified that justified the new protective structure.

In Brazil, a similar method was adopted by the Supreme Court in 2020, when the Referendum in a Precautionary Measure in the ADIs 6.387, 6.388, 6.389, 6.390, and 6.393 was judged. On this occasion, the fundamental right to data protection was recognized as an implicit right in the Brazilian Constitution in the face of the new risks of the information society.

In his ruling, Justice Gilmar Mendes expressly recognizes the need for new parameters of protection for fundamental rights when the system is confronted by risks of new technologies:

> That is why, far beyond the mere debate about communication secrecy, this Court must recognize that the legal discipline of processing and using information ends up affecting the system of protection of individual guarantees as a whole. The contemporary factual framework must be internalized in the reading and application of the Federal Constitution of 1988. In fact, I would dare to say that the idea that the parameters of protection of fundamental rights should be permanently open to technological evolution has never been strange to constitutional jurisdiction.[3]

The need for a new contextualization of fundamental rights before the scenario of development of technological innovations, which the Justice recognized in his vote, is currently being addressed by legal researchers in the international scenario with a theory called "digital constitutionalism."

In this sense, the relationship between human dignity and neurotechnologies is already recognized by the risks of loss of autonomy and personal identity.[4]

The constitutional debate, in the international scenario, is seeking mechanisms so that constitutional protections are effective against a global scenario of accelerated advances in the development of technologies. Discussions about neurorights are reflections of the concern on new risks to human beings and the need for legal protections.

Hoffman-Riem's primary focus is on the interplay between technological and social innovation, emphasizing how legal frameworks are adapted by social innovation to ensure the continued relevance of its principles in contexts that were not initially anticipated. This adaptation is considered a fundamental aspect of a normative-constitutional order that can adjust to new and unforeseen circumstances. Indeed, there are many norms whose text has remained unchanged, but which have been applied differently over time, taking into account the conditions of each era. The important thing is that normative principles, as an expression of the legislator's will, continue to serve as a reference and are implemented, even if socioeconomic or technological conditions change.[5]

---

[3] Supremo Tribunal Federal (2020), p. 101.

[4] See Bublitz (2013).

[5] Hoffmann-Riem (2015), pp. 41–42.

Therefore, interpretative adaptations are considered legitimate to uphold constitutional values and principles when confronted with changes resulting from technological progress.

This idea, that fundamental rights demand developments in the face of technological evolution, accompanies what is currently being called "Digital Constitutionalism," a field of constitutional thought development that seeks to develop a system of regulations for the protection of fundamental rights in the virtual environment. It draws attention to the criticisms about the multiplicity of uses for the term "digital constitutionalism," and that would not be supported by a constitutional epistemology.[6]

The conception in question is the subject of debate in the Brazilian context by Gilmar Ferreira Mendes and Victor Fernandes, who explore a method of constitutional control that provides principles and values as normative parameters. The approach represents a practical manifestation of digital constitutionalism, suggesting that constitutional adaptations to new technologies should be guided by fundamental constitutional principles, preserving the essence of the original constitutional legislator while responding to the challenges posed by technological innovation.[7] Two authors are cited as exponents in the elaboration of digital constitutionalism: Giovani De Gregório and Edoardo Celeste.

For Giovani De Gregório, digital constitutionalism is a constitutional approach adopted after overcoming a liberal conception in the European Union.[8] This change comes from a new form of private power in the digital environment, as a result of the concentration of data processing power and content organization, demanding that contemporary constitutionalism not only seek to create protections against threats from public powers, but also adapt to this new scenario of powers.

Through a historical analysis of European constitutionalism, he identified three distinct stages in relation to the regulation of new technologies.[9]

The first phase is called digital liberalism because it removes the need for state regulation for the online environment, advocating self-regulation of the market and online communities. The second phase is recognized as the phase of judicial activism, whose main figure was the Court of Justice of the European Union due to the recognition of fundamental rights in the Charter of Fundamental Rights of the European Union and the Treaty of Lisbon, resulting in a change of the economic view on relations in digital environments to a rights view. Finally, digital constitutionalism, the current phase, is characterized by an advance of the judicial gaze on the risks that arise as a result of the accumulation of powers by large corporations, which can rival states and are based on the development and provision of new technologies. For the author, this phase demands that the classic constitutional approach

---

[6] On the criticisms, see Pereira and Keller (2022).

[7] Mendes and Fernandes (2020), p. 10.

[8] De Gregorio (2021), p. 42.

[9] De Gregorio (2021), p. 43.

of limiting the powers of a public entity before individuals be expanded to protection against private agents in the technological context.

Edoardo Celeste shares with De Gregório the view that digital constitutionalism is a way to explain new constitutional reactions in the face of changes in the constitutional balance. Such changes are due to new private agents with powers that result from their technological control.[10]

The first point in favor of the systematization of digital constitutionalism are the three changes caused by new digital technologies in the constitutional ecosystem.

The first is the amplification of the exercise of fundamental rights by individuals as a result of how technology expands the possibility of information transfer, which results in a reinforcement of the right to freedom of expression, religious freedom, among others. The second change is the amplification of risks to fundamental rights, as a result of the first movement of amplification mentioned, as the abuse of these freedoms can configure new ways to threaten these rights. Finally, the third change reflects on the balance of powers in the constitutional ecosystem, given that large corporations with global reach begin to have control over fundamental rights, as these private agents begin to play a central role in the daily lives of people around the world.

In contrast to these changes, the researcher lists three normative reactions[11] to limit the effects of this scenario.

The first is the structuring of norms that begin to guarantee, in the digital environment, the exercise of existing fundamental rights. The second reaction is the construction of norms to limit the violation of fundamental rights. The author expressly cites the emergence of personal data protection. Here we can find a legal argument in favor of the structure of neurorights already described in Chap. 2. Finally, the last category of norms aims to restore the balance of constitutional powers between the involved agents: states, citizens, and corporations.

Thus, one can argue, according to Celeste, that constitutionalism digital, as a development of contemporary constitutionalism, requires the creation of normative countermeasures to confront changes in the constitutional balance caused by the advancement of digital technology, while also offering the ideals, values, and principles that guide such measures.[12]

Considering the theoretical context presented, digital constitutionalism would be an ideology[13] that establishes and guarantees a normative system for the protection of fundamental rights in digital environments. This is achieved through values and principles that permeate, guide, and inform the constitutionalization of the digital environment.

This constitutionalization of the digital environment is the process of normative production of reactions to changes in the balance of the legal system, which are

[10] Celeste (2019), p. 3.

[11] Celeste (2019), pp. 5–6.

[12] Celeste (2019), pp. 5–6.

[13] Celeste (2019), p. 13.

produced by digital technologies and occur in two dimensions: at the national level, through constitutional and infraconstitutional rules, and at the transnational level, through treaties and decisions of international courts or definitions of international private agents.

In the context of neurorights proposals, we can see that Celeste's systematization helps to identify how they should be analyzed, given the current constitutional moment: manifestations of normative reactions in the face of a scenario that neurotechnologies can influence, either as amplifiers of the exercise or threats to fundamental rights, or as a means of controlling constitutional powers by private entities.

It is important to note that, in the face of Celeste's system, neurorights can be seen as principles and normative values for the structuring of specific norms, and not necessarily as new rights.

This allows for an approach to analyze the elements identified in the neurorights proposals in relation to the construction of protection for human dignity, encompassing not only new elements that are outside of current protection, but also manifestations of the protection of dignity in the context of risks arising from the use of neurotechnologies.

4.2 The Protection of Human Dignity as a General Principle of Constitutional Protection

The conception of the existence of human dignity as a constitutional-legal dimension is the result of a historical construction that dates back to Antiquity and goes through religious and philosophical contributions.

This section will carry out a historical passage, from the time when dignity was seen as an abstract notion until it gained its recognition as a constitutional value. However, there is no intention to be exhaustive, considering that the focus of the work is different. In this sense, human dignity has a role in contemporary constitutionalism that should guide the interpretation of fundamental rights, demanding an understanding of its origin and evolution, especially with the Kantian construction.

First, a historical line of the main contributions of philosophers and religious ideologies to the notion of dignity will be traced.

Then, how this abstract notion was concretized as a legal concept will be presented, trying to identify the content of this principle and its dimensions.

Subsequently, human dignity as a fundamental norm in the Brazilian constitutional framework will be studied, identifying its role as a value, principle, and rule and its relationship with fundamental rights. In this section, the role of the Supreme Federal Court in its concretization will be presented.

4.2.1 Human Dignity as an Abstract Value

The historical construction of the concept of human dignity as an abstract value has, at least, 2000 years of contributions, with philosophical, religious, and political origins. Therefore, it would be impossible to exhaust in this work all this journey. However, it is possible to identify some of the milestones that are present to this day in the content of the concretization of such a concept.[14]

From religious origins, it is possible to quote biblical references that attribute to the human being the image and likeness of God, reinforcing that every individual has value by their very nature of being human. This religious foundation places the human species above all other animal species on Earth, reinforcing the impossibility of reducing a person to degrading conditions.

During the period of Classical Antiquity, dignity was regarded as an attribute associated with an individual's political and social status. For instance, in ancient Greece, free citizens were endowed with isonomic privileges and equal rights of speech in the agora. However, legal systems also recognized forms of subjugation, such as slavery and the subjugation of women.

The philosophical underpinnings of this phenomenon can be traced back to a philosophical discourse within Greek thought. This discourse posited the coexistence of the conception of equality with the notion of natural inequality among individuals. This was regarded as a legitimate justification for discrimination. This would be recognized even by Aristotle,[15] when he justifies slavery and the subjugation of women to men with a supposed division of human beings between those born with the dominion of the soul over the body and the control of instincts by reason, and those who are born without these qualities, slaves, and women.

In Ancient Rome, on the other hand, dignity was part of the concept of *dignitas*,[16] which was a value conditioned to social position and the degree of social recognition, which allowed the modulation of dignity for people more or less worthy.

It was in Cicero[17] that the conception of a universal dignity as a characteristic of all people was structured, in contrast to other animal species. It is precisely from this philosopher's conception that a moral sense of dignity is extracted, which includes personal virtues, merit, integrity, among others, and a socio-political sense, which covers the social and political position of individuals.

In the Middle Ages, the conception was taken up by Saint Thomas Aquinas as dignity both for the creation of man in the image of God and for free will, which guarantees the individual's capacity for self-determination.[18] However, even though he had this perspective, he recognized a hierarchical order among men, based on divine predeterminations.

---

[14] On the construction of the concept of human dignity in antiquity, see Sarmento (2016).

[15] Sarmento (2016), p. 118.

[16] Sarmento (2016), p. 30.

[17] Sarmento (2016), p. 31.

[18] Sarmento (2016), p. 31.

In the Renaissance, a period marked by philosophical changes that placed man at the center of various fields of knowledge, Giovanni Pico della Mirandola is cited as a thinker who initiates a concept of dignity linked to individual autonomy and attributable to all individuals, regardless of class or social position.[19] They could make use of it on their own initiative, even though, from this view, the possibility of rights guaranteed equally to all was not extracted.

The next temporal leap that brought changes to the concept of dignity was the Enlightenment, marked by the construction of egalitarian ideals and by the processes of secularization of the state. As the main figure of the period, Kant's conception of human dignity is debated to this day in the structuring of fundamental rights of constitutional charters, being mentioned in decisions of the Supreme Federal Court of Brazil that will be studied below.

In this sense, Jeremy Waldron attributes to the Kantian concept of dignity an extensive concept, which can be dimensioned to an individual as an almost value, not in the molds of "good" or "bad," but a value related to a normative type.[20]

In the legal constructions of Modernity, the generalization of fundamental rights, which is a consequence of a vision of the human being as the center of relations, ended up distancing the conception of dignity as an attribute linked to social recognition or economic power, with the conception of the individual as an end in itself at the starting point of the discussion in the constitutional field.[21]

4.2.2 The Realization of Dignity in the Face of Constitutional Practice

The exact content of dignity as an abstract value is difficult to define, especially considering the attempts to realize it in the constitutional sphere. This difficulty resulted in a debate that continues to this day in Law, Philosophy, and Politics.

In the context presented, Sarmento (2016) attributes to Kantian philosophy some direct effects on the realization of human dignity in constitutional charters that followed the philosopher's conception.[22] One of the characteristics that marked the construction of these constitutional charters is the fact that Kant's philosophy aimed at an abstract and disembodied human being. The dignity of an abstract human being, in conjunction with other political and economic elements of the nineteenth century, generated constitutional charters whose liberal ideals protected economic freedoms, and not existential ones. Self-determination as an element of dignity, for

[19] Sarmento (2016), p. 33.

[20] Waldron (2007), p. 11.

[21] Sarmento (2016), p. 109.

[22] On the abstract conception of human dignity in Kantian philosophy and its effects on the construction of fundamental rights in Modernity, see Sarmento (2016), Chapter 1.6—From the abstract individual to the concrete person.

example, was adopted, in this context, as a source of obligations linked to the private and patrimonial sphere, resulting in the state's abstention from the protection of existential minimums.

This period, which coincided with revolutions in the iron and coal industry, was marked by a degradation of the value of the human being in relation to capital and property, increasing poverty and lack of access to basic elements for human life.

The abstract view of the human being, culminated in an increase in misery and a true objectification of the human being in the capital machine, in addition to impacting the economy and social relations, caused dissatisfaction that drove the labor revolutions in Europe and other countries, resulting, in some countries, even in the rupture of the constitutional order (as in the Russian Revolution of 1917, for example).

The crises of the liberal states originated from revolutionary movements against the oppressions of capital, allowing us to observe the beginning of a legal realization of human dignity as a value to be protected in the face of real human beings, in contrast to the abstract view that was present until this historical moment.

It is in this context that dignity as an inherent quality of the human being encountered conceptual difficulties to be realized as a legal-normative condition capable of protecting human existence.

Therefore, the answer to the question of how to achieve realization necessitated an examination of dignity from a standpoint of concrete factual situations, rather than from the perspective of generic terms. This approach would allow for a more nuanced understanding of how a given society actualizes the dignity of the human person.[23]

This realization approach moves away from a fixed conception in favor of a pragmatic solution aimed at the purpose of protecting human dignity: protections against violations to it are sought. For this, minimum and basic contours were built by doctrine and jurisprudence, with an emphasis on identifying cases of aggression against them and, from concrete situations, developing their content.

However, it was only after World War II[24] that it became evident to the vast majority of countries the impossibility of ignoring the need for provisions on human dignity as a principle and value to be realized in the application of legal orders. The result was a proliferation of the principle in various countries and multilateral treaties between states.

On the international scene, the proclamation of the dignity of the human person began to appear in the following documents: UN Charter (1945), Universal Declaration of Human Rights (1948), International Convention for the Elimination of All Forms of Racial Discrimination (1965), the Covenant on Civil and Political Rights (1966), the Covenant on Economic, Social and Cultural Rights (1966), the Inter-American Convention on Human Rights (1978), the Convention on the Elimination of All Forms of Discrimination against Women (1979), the African Charter on Human and Peoples' Rights (1981), the Convention against Torture and

[23] Mendes (2013), p. 90.

[24] Mendes (2013), p. 85.

Other Cruel, Inhuman and Degrading Treatment (1984), the Convention on the Rights of the Child (1989), the Charter of Fundamental Rights of the European Union (2000), and the Convention on the Rights of Persons with Disabilities (2007).[25]

In the context of national states, the German Basic Law of 1949 was responsible for effectively transforming human dignity, which was a philosophical value, into a constitutional textual provision in Art. 1, I: "The dignity of the human person is inviolable. To observe and protect it is the duty of all state powers." Following the example, Portugal (1976), Spain (1978), and Brazil (1988) began to provide for the dignity of the human person in a special hierarchy.[26]

From this realization, Sarlet characterizes[27] dignity with a multidimensional character, considering: ontological dimension, historical-cultural dimension, double negative and positive dimension, in addition to a double objective and subjective dimension as a principle and norm that underpins fundamental rights.

4.2.3 *Human Dignity in the Brazilian Constitutional System*

The Federal Constitution of Brazil provides for the dignity of the person as the foundation of the democratic rule of law (Art. 1, clause III),[28] listing it as the purpose of the economic order to ensure a dignified existence for all Brazilians (Art. 170)[29] and establishing it as one of the foundations for family planning (Art. 226, §7),[30] as well as one of the duties of guarantee for the family, society, and state towards children and adolescents (Art. 227)[31] and towards the elderly (Art. 230).[32]

---

[25] Sarmento (2016), pp. 54–55.

[26] Mendes (2013), p. 85.

[27] Sarlet (2020), pp. 16–24.

[28] Article 1. The Federative Republic of Brazil, formed by the indissoluble union of the States and Municipalities and the Federal District, constitutes itself as a Democratic State of Law and has as its foundations: III—The dignity of the human person.

[29] Article 170. The economic order, based on the valorization of human work and free initiative, aims to ensure everyone a dignified existence, according to the dictates of social justice, observing the following principles:

[30] Article 226. The family, the basis of society, has special protection from the State. § 7 Based on the principles of human dignity and responsible parenthood, family planning is the free decision of the couple, it is up to the State to provide educational and scientific resources for the exercise of this right, any coercive form by official or private institutions is prohibited.

[31] Article 227. It is the duty of the family, society and the State to ensure children, adolescents and young people, with absolute priority, the right to life, health, food, education, leisure, professionalization, culture, dignity, respect, freedom and family and community coexistence, in addition to keeping them safe from all forms of negligence, discrimination, exploitation, violence, cruelty and oppression.

[32] Article 230. The family, society and the State have the duty to support the elderly, ensuring their participation in the community, defending their dignity and well-being and guaranteeing them the right to life.

Our Constitution adopted an anthropological premise of dignity, focusing on concrete individuals, with physical bodies and imbued with rationality and feelings, an end in themselves, and who live in society with the need to have their autonomy respected. Thus, the normative sense of dignity goes through a constitutional conception of the centrality of the person as the end of the legal order and not as a means, mainly with the provision of fundamental rights in Art. 5 and their quality as an eternity clause in Art. 60, §4, clause IV; Starting from an interpretation in the light of critical morality, "the person, in this sense, has an intrinsic value, and cannot to be instrumentalized. This applies to absolutely every single person, regardless of their social status, or the heroic or heinous acts they may have committed: everyone has equal dignity."[33]

The person to whom the Constitution refers, therefore, is a subject with capacity and autonomy guaranteed through individual, public, and private freedoms, and who is endowed with rationality, emotion, and social relations.[34]

Once the concept of person that the Constitution addresses is established, starting from a systemic interpretative reading of normative provisions and interpretations of the Federal Supreme Court, it is possible to delve into the normative role that dignity plays in Brazil.

This concept is not just a fundamental right, but a principle endowed with axiological and normative value, with the role of foundation of norms defining rights, guarantees, and fundamental duties, being, at the same time, a supreme constitutional principle and a fundamental right.[35]

As the foundation of the legal order and the community, human dignity assumes, in Brazil, a role equally recognized by international communities in instruments such as the International Covenant on Civil and Political Rights and the International Covenant on Economic, Social and Cultural Rights of the United Nations. This foundational role has two dimensions: that of moral legitimacy, which gives foundation to the state and the legal order, existing because of the person; and that of a hermeneutic role, should guide interpretative processes in the legal field, considering that it is present in the constitutionally provided fundamental rights of freedoms, political, social cultural and transindividual. It emphasizes that, due to its foundational role, it should also be observed in the interpretative role of all norms defining the state and private relations.

When determining the relative weight of conflicting constitutional interests, this criterion is employed to assign greater weight to legal goods that are most closely associated with the protection and promotion of dignity.[36]

---

[33] On the functions described here, see Sarmento (2016), Chapter 2.3—The functions of the principle of human dignity.

[34] Sarmento (2016), p. 77.

[35] Mendes (2013), p. 90.

[36] Sarmento (2016), p. 81.

Its function as a limiter of fundamental rights has a dual dimension: limitation due to the protection of the human dignity of a third party and restriction of the right of the holder in the name of the guardianship of his own dignity.

It should also serve as an interpretative tool in private relations, being endowed with horizontal efficacy among individuals to protect the interests that are closest to dignity in cases of conflict. Thus, as dignity radiates to the private sphere, it should also be observed by the administrative acts of the state in its various species and spheres.

Finally, perhaps its most important function for the central hypothesis of the study of neurorights, is to be a source of criteria for identification of fundamental rights not expressly provided for in the constitutional list of Art. 5°, since its role is to integrate the content of fundamental rights or, at least, a minimum projection of these rights.[37]

Therefore, by assuming an umbilical character with fundamental rights, dignity also becomes a criterion for identifying these rights. This is not expressed in the constitution, and it enables the structuring of defenses against serious threats to injuries due to gaps. It also plays a role in guaranteeing constitutional openness to new social contexts.

After establishing the significance of the dignity of the human person within the constitutional order, as delineated in the preceding item, it is imperative to endeavor to delineate a minimum content and fundamental ordering principle for Brazil.

The content of this principle can be divided into four elements: intrinsic value of the person, autonomy, existential minimum, and recognition.[38]

The first element, intrinsic value of the person,[39] originates in the use of senses in which dignity is normally employed: status, virtue, and intrinsic value. Dignity as status is the current use, which comes from the beginnings of conceptions about the term, in relation to the social aspect that the individual assumes or, in other terms, his recognition before society. As a virtue, dignity is manifested as dignified conduct, translated through actions and postures valued in the social environment. The third sense in which dignity is adopted corresponds to its ontological character,[40] related to the very existence as a human being, independent of other attributes.

This last sense originates in the Kantian idea that people are ends in themselves, not mere instruments that can be used for the purposes of third parties or society itself, should be treated as rational subjects, capable of making choices and self-determining, being prohibited from constraints, manipulations, or lies.

This dimension of dignity was adopted in Brazil by the Supreme Federal Court in, at least, two important cases that have a relationship with biological aspects of human beings, which are related to the theme of neurorights.

---

[37] Sarlet (2020), p. 28.

[38] Sarmento (2016), p. 92.

[39] See Sarmento (2016), Chapter 3—The Intrinsic Value of the Person.

[40] See Sarlet (2020), p. 23, footnote 27.

The first was the ADI n°3510/DF, reported by Justice Ayres Britto, which dealt with research on embryonic stem cells by questioning the Biosafety Law, Law n° 11.105/05, in the limitation of Art. 5°: Embryos produced by in vitro fertilization, not used in the procedure, which prove to be unviable or frozen for more than 3 years, both hypotheses with the consent of the parents. The argument for unconstitutionality was the interpretation that the legislator treated the embryo as a means of research, and not as an end in itself.

This interpretation was dismissed on the grounds, according to the ruling of Justice Celso de Mello, that:

> There is no offense to the right to life and the dignity of the person human, as research with embryonic stem cells (biologically unviable or for the purposes for which they are intended) signifies the celebration of life and hope for those who are on the margins of the exercise concrete and inalienable rights to happiness and living with dignity.[41]

It was recognized the non-existence of violation of dignity because: (1) the limitations guarantee the use of unviable embryos, which are not endowed with the capacity for development into human persons, and (2) there is the realization of the dignity of those people who can benefit from new therapies resulting from research with embryos.

At this point, it is possible to establish a parallel between the dignity in the use of neurotechnologies that enable essential aspects of the human person and the use of devices that guarantee the return of lost senses or physical limitations. Also, dignity can be imposed as a prohibition of the negative effects of neurotechnologies for the collection of statistical data that result in economic gains for companies (a human person as a mere test tool).

Another more recent judgment was the ADPF n°54, which concerns the abortion of anencephalic fetuses. The prohibition of abortion in Articles 124, 125, and 128 of the Penal Code was questioned in the case of fetuses with unviability of life. Sarmento (2016) highlights the ruling of Justice Marco Aurelio:

> The woman, therefore, must be treated as an end in herself, and not, from a utilitarian perspective, as an instrument for the generation of organs and subsequent donation. Even if the organs of anencephalics were necessary to save other lives - a premise that is not confirmed, as will be seen -, she could not be compelled, on the basis of solidarity, to carry the pregnancy to term, imposing all kinds of suffering on her. Otherwise, she would be seen as a simple object, in violation of the human condition.[42]

By recognizing dignity as an intrinsic quality, which prevents the instrumentalization of the human being, even for the benefit of society, the vote helps in the realization of dignity in the face of the effects of neurotechnologies, even in situations where society benefits from them.

The points made earlier about the possible use of these technologies in socially undesirable individuals (such as rapists, serial killers, etc.), aiming to modify these behaviors of their personalities, therefore, can be understood as an attack on human

[41] Supremo Tribunal Federal (2010), p. 136.

[42] Supremo Tribunal Federal (2012), p. 52.

dignity in its element of intrinsic value, if it occurs without the consent of the individuals.

The second element of dignity dealt with by Sarmento is autonomy,[43] with two possible meanings: self-determination as the right to make choices, as long as without harm to others' rights, and as a limitation of autonomy to prevent the voluntary submission to situations of indignity. It is the conception of freedom as a constitutional-legal manifestation.

It is divided into public and private, with the former referring to the possibility of the individual making their own choices and the latter addressing the power to participate in political decisions in social life.

Autonomy deals with the possibility of making decisions and choosing what to desire for your personal life, regardless of the motivation for such decisions (emotional, rational, etc.), there being limits to choices and actions given the possibility of harm to third parties.

The realization of dignity as autonomy was the subject of the aforementioned case of abortion of anencephalic fetuses, appearing in two passages of votes. The first of them, from Justice Marco Aurélio, says he is: "in play the right of the woman to self-determine, to choose, to act according to her own will." The second, from Justice Joaquim Barbosa, determined that "procreation, gestation, in short, reproductive rights are inseparable components of the fundamental right to freedom and the principle of personal self-determination, particularly of the woman."

Also, the element of autonomy as belonging to dignity and related to the Kantian perspective was mentioned in the ruling of Justice Luiz Fux in a decision that dealt with the constitutionality of stable homoaffective union, ADI n°4277 and ADPF n°132: "it is up to the State to ensure (…) that each one can lead his life autonomously, according to his own designs and that sexual orientation does not constitute an obstacle to the pursuit of personal objectives (…) this order of ideas refers to the issue of private autonomy of individuals, conceived in a Kantian perspective, as the center of the dignity of the human person."

Self-determination can also be public, having instrumental importance because the participation of individuals in political life can promote improvements in living conditions and, also, can have value constitutive, with the individual being an agent in the construction of the state and not only a recipient of norms.

At this point, concerns about the use of neurotechnologies for manipulations that affect democracy have been raised before, regarding the use of personal data by Cambridge Analytica in the United States presidential elections in 2016, as these new devices can reach levels of manipulation that take away the public autonomy of individuals to decide on life in society.

The third element of dignity is the recognition of the minimum essential for the person in social life, which is closely linked to the principles of equality and solidarity.[44]

---

[43] See Sarmento (2016), Chapter 4—Autonomy.

[44] See Sarmento (2016), Chapter 5—The Existential Minimum.

As normative content, recognition goes through uniform respect for identity, and violations of this element of dignity include state or private practices, conscious or not, that may endanger the identity of others in a way that brings negative feelings. With this, it is marked by the characteristic of respect for individual identity to guarantee a dignified life in society, especially to protect racial, ethnic, religious minorities, etc.

It has a negative character by prohibiting practices that disrespect the individual identity, and a positive character by demanding that the state adopt measures to combat these practices. This element can generate tension with the element of self-determination, since it can end due to this element and lead to interferences in the subjectivity of individual conceptions, and it should be the role of the state to decide on the degree of recognition, in relation to individual freedoms, guided by proportionality.

According to the author under study, there are three normative factual contexts in which recognition manifests itself: protection of universal rights, guarantee of special rights to members of specific groups, and protection of traditional peoples.

The first manifestation seeks to ensure that minority groups are not excluded from universal rights, an example being the possibility of recognition of same-sex unions, given that, once the stable union is recognized, they are still denied civil marriage.

The second manifestation is about establishing specific protections to correct structural and historical injustices, adapt norms and practices that were established by majority groups and enable the survival of minority cultures, which harmonize with a general notion of equality. An example of realization is Law No. 13.146/15, the Statute of the Person with Disabilities.

The last manifestation concerns the protection of cultural differences of traditional peoples, focused on ensuring the existence and reproduction of original cultures. This realization manifests itself in the right to lands traditionally occupied by Indigenous people (Art. 231 of the CF) and quilombolas (Art. 68, ADCT).

4.3 The Relationship Between Human Dignity and Fundamental Rights in the Analysis of Neurorights Proposals

Once the normative content of human dignity as a fundamental guiding principle of the legal system in Brazil has been studied, for the deepening of the analysis with the proposals of neurorights at the constitutional level, we will examine the relationship between the theory of fundamental rights and human dignity.

There is a double link between these two figures: dignity has a role of direct and immediate foundation for fundamental rights, which are concretizations of this fundamental principle; it is also, ultimately, the content of these rights, albeit in different ways and intensity.

Thus, fundamental rights can be understood as the realization of various dimensions of human dignity, depending on the context of protection. For this reason, the constitutional opening for the list of fundamental rights provided for in Art. 5°, §2° of the Federal Constitution,[45] is justified, since it is possible to realize these dimensions in situations to be defined according to social and cultural needs.

As previously highlighted, one of the functions of human dignity is to be a criterion for recognition and delimitation of fundamental rights, considering that, at a certain level, every fundamental right must have a core of protection for dignity. It is common for concepts such as "human rights," "fundamental rights," and "protection of dignity" to arise when dealing with the topics set forth here.

The conceptual difference depends on the plane in which it is being dealt with. The expression "human rights" is widely used internationally, regardless of how each national state regulates the subject. On the other hand, "fundamental rights" usually refers to the rights enshrined in a specific constitution of a state. For this reason, this term has been preferred when addressing the protection of the human person in the context of public law, especially in relation to the action of state power. The expression "personality rights" refers to human attributes that require special protection in private relations, that is, in interactions between individuals, although they also have a constitutional basis and protection at the national and international level.[46]

Thus, personality rights are those that seek to protect human dignity in private relations and at the infraconstitutional level, while fundamental rights do the same at the constitutional level, and human rights at the international level. As mentioned above, neurorights theories call for international recognition as human rights, to then influence national constitutional orders.

Not all fundamental rights, even those intended for the human person, find correspondents in the protection of personality. On the other hand, all personality rights can find correspondents in the constitutional sphere as fundamental rights.[47]

Using these distinctions, it is worth mentioning that the analysis carried out will be dealt with at the constitutional level, specifically, within the context of the Constitution of Brazil. Below, a comparative analysis of the protective content of neurorights in relation to those already existing in the Brazilian constitutional order will be presented.

---

[45] § 2° The rights and guarantees expressed in this Constitution do not exclude others arising from the regime and principles adopted by it, or from international treaties in which the Federative Republic of Brazil is a party.

[46] Schreiber (2014), p. 13.

[47] Sarlet (2020), p. 194.

4.3.1 The Right to Free Development of Personality and Cognitive Freedom

As an introduction, it is necessary to reconstruct a brief history of personality rights in Brazil.

The first structures on the idea of a right to personality were associated with a concept of "aggregation of attributes particular to the human condition," or special human manifestations whose protection is necessary for the development of people. These manifestations occurred through judicial activity and were later shifted to discussions in the field of private law, influenced by the civil codifications of Italy, Germany, and Portugal.[48]

In the 1960s, there was a "social turn," led by jurists based on European movements for the establishment of a welfare state, resulting in the fall of legal individualism and the role of law socialization. This turn was, in turn, subverted by the Military Coup of 1964, which even eliminated fundamental rights through Institutional Act No. 5 of 1969.

With the Federal Constitution of 1988, the dignity of the human person was listed as a fundamental value and principle, as analyzed in the previous item, generating profound changes in the Brazilian legal system. However, unlike the Basic Law of Germany, which provides for the free development of personality[49] in Art. 2.1, alongside the dignity of the human person, the Brazilian Constitution does not make such a provision.

Even without express provision, there is consensus on the existence of a general clause of free development of personality, resulting from the constitutional text.[50] A reading of Art. 1°, clause I and III; Art. 3°, clause III, and Art. 5° *caput* and §2° opens the possibility of infraconstitutional elaborations through axiological parameters constitutionally disposed. It is from this context that arises, as the highest value by the legal system, a general clause of protection and promotion of the human person.[51]

The protection for the dignity of the human person has an element of elasticity, which allows safeguards against threats that are foreseen, or not, in the legal system, guaranteeing a right of freedom, which prevents interferences that harm the development of personality and personal self-determination. In the words of Sarlet:[52]

> It is possible to affirm that the general right of personality (or right to free development of personality) implies comprehensive protection against any form of violation of personality goods, whether they are, or not, expressly and directly recognized at the level of the constitution.

---

[48] Doneda and Zanatta (2022), p. 39.

[49] Article 2.1. Everyone has the right to the free development of their personality, as long as they do not violate the rights of others and do not violate the constitutional order or moral law.

[50] See Mendes (2013), Sarmento (2016), and Sarlet (2020).

[51] de Assis Zanini and Queiroz (2021), p. 18.

[52] Sarlet (2020), p. 196.

This general clause coexists with special personality rights, not only as a complementary or symbolic right, but as an autonomous fundamental right, which seeks to guarantee the formation and free development of personality in conjunction with the protection of individual freedom of action and the integral protection of the person, unlike special rights that seek specific protections.

The need for a general clause is due to rapid changes in societies and technologies, which end up generating difficulties for rights recognized only by positivation, demanding mechanisms that operate at a non-specific level.

The characteristics of this right are: (1) universality; (2) absolute character; (3) extrapatrimonial; and (4) unavailability; they are conferred both in the general right to personality and in special rights as well as in fundamental rights, with some particularities reserved.

Universality deals with the ownership of personality rights: as a consequence of the characterization of subjective rights, these rights are guaranteed to any existing and real individual. Thus, this category of rights accompanies every individual who is born alive until the moment of their death, with some protections of objective dimension being established for before life and after death.

The absolute character refers to the possibility of being opposable to all, whether private individuals or state entities, and the impossibility of total non-applicability of these rights when in collision with other fundamental rights or personality rights of third parties, should be applied the technique of solution through pondering.

They are extrapatrimonial because they relate to values, goods, or interests linked to subjectivity, even though the injuries may have patrimonial repercussions.

The unavailability depends on the specific case, since its patrimonial effects can be transferred, or these rights can be self-limited by the holder himself. A parallel can be drawn with certain qualifications of fundamental rights, such as the distinction between the absolute and irrevocable renunciation of the right itself and the partial and revocable disposition of aspects related to its exercise within the individual's purview.

To understand the right to free development of personality, besides the general clause, it is necessary to study the relationship of this right with the special rights of personality.

The general clause, as a rule, has subsidiary application to the special rights that, due to their positivation, may have limited effectiveness specifically regarding the scope of their application. This coexistence in the general character and the specific protections for particular goods expand and condition the protection of personality to the specific case, making the safeguards faster and placing the rights in a true situation of complementarity.[53]

The relationship of coexistence, however, has consequences that are subject to study that allow to aggregate similar views on the aspects of this relationship.

The pluralist current, more adopted in Brazil, understands that these various rights of personality exist independently of each other. One of the points of view of

[53] de Assis Zanini and Queiroz (2021), p. 14.

this current argument is that the special rights are open to new constructions, and another is that they are taxative.[54]

This approach substantiates the notion that the general clause of free development of personality, as it pertains to the establishment of guardianship founded on technological innovations, is subject to verification. This is particularly salient considering the potential damages that such innovations may inflict upon personal development.

Bublitz, when structuring a normative conception of the right to cognitive freedom, performs an analysis with some rights that are close to this new conception, but do not have an exact scope of application for interventions in mental processes. Analyzing the content and role of the right to free development of personality, this right can be compared to the concept of cognitive freedom as a legal concept such as: "*It is, or rather, should be the central legal principle guiding the regulation of neurotechnologies, guaranteeing the right to alter one's mental states with the help of neurotools as well as to refuse to do so.*"[55]

The comparison can be made as to the function of both, which consists in guaranteeing protection to the dimensions of the human being that are not directly protected by other rights. While the free development of personality protects spheres of personality to guarantee the development of the human being, cognitive freedom protects the right to choose whether or not to use neurotechnologies (including, here, those that are related to chemical substances), as a guarantee of control over mental activities manifested through neural data.

The comparison is also made by the author himself, who identifies the general right to personality as a right that has strong protections against actions by third parties to the "hard core" of personality and that, at the same time, guarantees weak protections for the elements that are aimed at the development of personality by the individual himself. He explains:[56]

> By contrast, actions by individuals to actively develop and transform their personalities are protected to a considerably lesser degree. In German law, this freedom is synonymous with the basic and easily limitable right to free action. For example, the ban on cannabis has been tested against this weak right only. In a constitutional challenge against its prohibition, the German Constitutional Court held that cannabis consumption falls under the right to develop one's personality, but only in this weaker form, hence it is limitable by reasonable public interests.

This difference in protection, exemplified with the consumption of *cannabis* in Germany, can be applied in the case of neurotechnologies, since the current factual scenario indicates that the scope of their use will be massified for entertainment purposes and that, therefore, they will not constitute activities that are part of the "hard core" of personality.

---

[54] de Assis Zanini and Queiroz (2021), p. 19.

[55] Bublitz (2013), p. 234.

[56] Bublitz (2013), p. 249.

This, therefore, makes cognitive freedom a general clause for the freedom of use, or non-use, of neurotechnologies, with protections to the threats that can arise from them in a degree similar to that of free development of personality.

Thus, we see two distinct scopes of application for the rights, but both originating in the protection of human dignity when specific constructions are lacking.

4.3.2 The Right to Physical and Psychic Integrity and Its Relations with Mental Integrity and Psychological Continuity

The Brazilian constitutional order of 1988 did not express a right to physical and psychic integrity. However, human dignity necessarily encompasses respect and protection for the physical and emotional (or psychic) integrity of the person in general. From this premise, it follows that, by example, the prohibition of the death penalty, torture, the application of corporal punishment, and even the use of human beings in scientific experiments.[57]

This construction is systematic, encompassing the scenario of international treaties of which Brazil is a signatory and provisions of the Federal Constitution itself. They are: (1) The UN Declaration of Human Rights, from 1948, which prohibits torture, cruel, inhuman or degrading treatment or punishment; (2) The International Covenant on Civil and Political Rights of 1966, which, in addition to the UN prohibitions, also prohibits the conduct of medical or scientific experiments without the consent of the holder; (3) The Convention against Torture and Other Cruel, Inhuman or Degrading Treatment or Punishment (1984); and (4) The American Convention on Human Rights of 1969, which establishes the right to respect for physical, psychic and moral integrity.[58]

In the context of the 1988 Constitution, it is first pointed out that the right to integrity seeks, unlike the right to life, to protect precisely from interventions that do not result in death or that do not pose an effective risk. The right to physical (bodily) and psychic integrity covers the protection of the external integrity of the person, that is, the bodily dimension in its biological aspect, as well as internal integrity, related to the functioning of the psychic sphere, including the sensitivity to pain and physical and emotional suffering.[59]

This construction is of subjective rights with a dual dimension: negative and positive.[60]

The negative dimension refers to a defense protection against aggressions and offenses to physical and psychic integrity in the biological and psychological senses,

[57] Sarlet (2006), p. 88.

[58] Sarlet (2023), p. 193.

[59] Sarlet (2023), p. 189.

[60] See Sarlet (2023), Chapter 4.2—The Right to Life.

being opposable to state and private entities. It is mitigated by the consent of the holders, which seeks to guarantee the freedom of decision against interventions, by accepting those that they understand to be better for themselves. The positive dimension has an interface with the right to health, aiming at state provisions that ensure the physical and psychic integrity, not only in a way to curb violations, but also against threats and risks.

As it is a realization of human dignity and a kind of personality right, the ownership is universal, covering all people from the moment of the beginning of life until death, with reservations of personality projections for cases of the unborn and protections related to the dead.

One of the problems is the identification of legitimate interventions and interventions that violate the right, since, as a rule, the holder can consent to actions that cause loss of integrity. This consent must be free, unequivocal, and informed to be able to remove the illegality of the intervention, allowing it to be carried out when, in exceptional cases, the holder is not able to consent, being the intervention is necessary for him. The interventions will also be legitimate when based on acts: of the legislator, administrative (power of policy), and judicial, without necessarily existing consent of the holder, as in the case of compulsory extraction of biological materials for the production of judicial evidence or medical treatments.

The right to physical and psychological integrity is the same that gives rise to the right to mental integrity of Ienca and Andorno, there being a difference, according to the authors, in the scope of protection: while the first is considered a safeguard for health, the second is an expansion to avoid damage to neural processes.

This comparison is made by them:

> We propose to fill this normative gap by calling for a reconceptualization of the right to mental integrity. In fact, while the ECHR and the EU Charter of Fundamental Rights consider mental integrity as a right to mental-health, pendant of the right of physical integrity understood as physical health, a more complex dimension of mental integrity is elicited by neurotechnology. Mental integrity in this broader sense should not only guarantee the right of individuals with mental conditions to access mental health schemes and receivepsychiatric treatment or support wherever needed. In addition to that, it should also guarantee the right of all individuals to protect their mental dimension from potential harm.[61]

While mental integrity is based on safeguarding against harm, psychological continuity has as its scope of protection situations where there are no damages, but rather changes in personality.

Even though the authors point out these differences, it is possible to verify that in Ingo Sarlet's conceptualization, the scope of protection of physical and psychic integrity reaches both the organs that make up the central and peripheral nervous system, as well as the psychic sphere. Not being limited to the medical sphere and health protection, such conceptualization applies to any context.

Thus, it seems that, before the current construction of Brazilian authors, the scope of protection of these kinds of neurorights would already be covered in Brazil.

[61] Andorno and Ienca (2017), p. 18.

4.3.3 *The Right to Privacy and Personal Data Protection and the Right to Neuroprivacy*

The origins of the right to privacy and threats of social and technological changes are found in the article "*The Right to Privacy*" by Warren and Brandeis, which sought to identify a right to privacy based on English court rulings.

The article identified the following limits: (1) this right does not prevent the disclosure of what is of general interest; (2) it does not prohibit the communication of everything private, there being hypotheses for legal communication; (3) there is no compensation if the intrusion occurs by means that do not cause damage; (4) the victim's consent removes the violation of the right; (5) the claim of truthfulness of what is disclosed does not remove the illegality of the violation; and (6) intent is not necessary for the verification of the violation.[62]

These limits characterize an individualistic quality that reinforces the idea of *the right to be left alone* and its dimension of negative right, which demands absolute abstention from the state in the private lives of citizens. This individualistic character is still present in contemporary law, as it manifests as essential for a democratic society and is "a fundamental prerequisite for the exercise of various other fundamental freedoms."[63]

This individualistic conception, which guaranteed its exercise to privileged elites, gained new forms with the transformation of the liberal state into a welfare state, and the consequent modifications of relations between citizens-state. Here, it is important to refer to the change in the vision of human dignity from a mere abstract value to the demands for concretizations that safeguarded concrete human beings and not only ideal and abstract ones, largely after World War II.

Concurrently, technology evolved, and the demand for information flow by states and private entities increased. In the case of the state, the welfare state demanded greater collection of information from its citizens for better knowledge and social control. As for private entities, the value of information as an economic substrate began to be considerable, since the costs of collecting and processing information began to decline along with the cost of access to new technologies.

As the power structure underwent modifications with the informational relationship that was established, the notion of privacy demanded transformations for more specific concretizations, something that is verified from the 1970s, with specific legislation and judicial decisions that focused on personal data.

In Brazil, on the other hand, the right to privacy is not expressly provided for in the Federal Constitution.

Justice Gilmar Mendes, when judging ADI n° 6649 and the ADPF n° 695 in the Federal Supreme Court,[64] explains how this individualistic and negative conception came to Brazil through the work of Professor Tércio Ferraz and appeared in Brazilian

---

[62] Mendes (2014), p. 28.

[63] Doneda (2019), p. 31.

[64] Supremo Tribunal Federal (2022), p. 44.

courts as a restrictive jurisprudential guideline regarding the scope of the protection of Article 5, clause XII, of the Federal Constitution, undergoing changes in the twentieth century. The technique adopted by the constitutional legislator was to recognize dimensions that make up the right to privacy and intimacy, as well as honor and image. Even though they refer to different spheres, with intimacy being considered more restricted (it comprises the personal and family life of the holders) than privacy (which, in turn, comprises professional, commercial relationships, etc.), both must be analyzed together, considering the reality of current social contexts that often end up confusing these different spheres. This approach was consolidated in clauses X,[65] XI,[66] and XII[67] of the constitutional text, covering various specific aspects of the protection of private life and intimacy, in addition to different spheres of private life. From a formal criterion, two scopes of protection are identified: a more protected one, which does not admit state interventions, and a more open one, related to the social life of the individual, which is susceptible to intervention.[68]

However, considering the thin line of what can or cannot be considered of interest to the state to justify interventions, Ingo Sarlet argues that a material criterion should be adopted for the scope of protection of privacy: "what is considered reserved and unavailable to the interest of the State and third parties is that which is in accordance with the current social guidelines."[69]

The right to privacy is a subjective right, characterized by being a defensive right that restricts the action of the state and third parties in its scope of protection, and is also characterized as a right of freedom, as it seeks to ensure the individual's right to live life according to their desires, feelings, and convictions, as long as this does not extend to the possibility of causing harm to others. Its objective aspect refers to the obligation of the state to create mechanisms for protection against private individuals.

The scope and limitations of privacy are linked to its flexibility, "as a result of the configuration of the values of a certain society."[70] As an example of this flexibility, the construction of the United States on *privacy* as a fundamental right that guarantees human dignity was used as justification for the right to abortion in the Roe v. Wade case of 1973 by the country's Supreme Court.

---

[65] X—The intimacy, private life, honor and image of people are inviolable, with the right to compensation for material or moral damage resulting from their violation.

[66] XI—The home is the inviolable asylum of the individual, no one can penetrate it without the consent of the resident, except in case of flagrant crime or disaster, or to provide aid, or, during the day, by judicial order.

[67] The secrecy of correspondence and of telegraphic, data and telephone communications is inviolable, except, in the latter case, by court order, in the cases and in the form that the law establishes for the purposes of criminal investigation or criminal procedural instruction.

[68] Sarlet (2023), p. 204.

[69] Sarlet (2023), p. 205.

[70] Doneda and Zanatta (2022), p. 108.

Like other rights based on human dignity and because it has been rooted as a fundamental right, privacy is also subject to limitations[71] when confronted with other fundamental rights or relevant constitutional goods, each situation must be weighed in the concrete case for the resolution of conflicts, which returns to the unitary character of human dignity that must be interpreted according to each legal system.

As dimensions of privacy in the Brazilian legal system, we have: fiscal and banking secrecy, inviolability of domicile, inviolability of correspondence and secrecy of communications, and the protection of personal data.

Even if not constitutionally expressed, fiscal and banking secrecy is considered by doctrine and jurisprudence as of constitutional character. This right protects the information contained in fiscal and banking records, even if they reveal personal information.

Finally, the last dimension of privacy that will be studied and has the greatest interaction with the proposal of neuroprivacy is the protection of personal data. As exposed in the historical evolution that opened this section, data protection was a topic that emerged in the 1970s, after political, philosophical, and technological changes that brought difficulties to the individualistic conception of privacy.

In Brazil, this topic was treated in a sectoral manner and with little depth until the advent of the General Data Protection Law (Law No. 13.709/18). The following provisions are commonly cited that mention data protection superficially: Consumer Protection Code (Law No. 8.078/90), Civil Code (Law No. 10.406/02), Positive Registration Law (Law No. 12.414/11), decree of the Consumer Service (Decree No. 6.523/08), decree for a single register of federal government social programs (Decree No. 6.135/07), and decree of the Annual Education Census (Decree No. 6.425/08).[72]

However, the General Data Protection Law brought a comprehensive data protection system to the Brazilian legal system, prompting discussions in various legal spheres.

At the constitutional discussion level, the Federal Supreme Court recognized an implicit fundamental right to personal data protection in the referendum in Preliminary Injunction in ADIS 6.387, 6.388, 6.389, 6.390, and 6.393. The case questioned the constitutionality of Provisional Measure 954/20, which determined the sharing of data between telephone companies and the IBGE for the census due to the COVID-19 pandemic.

Some Judges expressly identified the origin of unconstitutionality in relation to the right to privacy, to human dignity, and to personality, as is the case with Justices Luis Fux and Gilmar Mendes.

On the constitutional origin of data protection in Brazil, the court identifies:[73]

---

71 Sarlet (2023), p. 206.

72 Mendes (2014), p. 141.

73 Supremo Tribunal Federal (2020), p. 109.

> The assertion of a fundamental right to privacy and to the protection of personal data derives, on the contrary, from an integrated understanding of the constitutional text based (i) on the fundamental right to the dignity of the human person, (ii) on the realization of the permanent commitment to renew the normative force of constitutional protection to intimacy (art. 5°, clause X, of the CF/88) in the face of the spread of new risks derived from technological advancement and also (iii) on the recognition of the centrality of Habeas Data as an instrument for the material protection of the right to informational self-determination;

According to the ruling, this fundamental right has the content of human dignity and the realization of protection to intimacy in the face of new normative force challenges, in addition to being based on the existence of *habeas data* as a constitutional instrument for the protection of informational self-determination.

The fundamental right to personal data protection has a dual scope of protection:[74] protection of the individual against threats to their personality resulting from data processing, and guarantee of data flow controls in society. In the subjective dimension, it is a defensive right, which guarantees the individual the cessation of intervention, as well as tools so that damages arising from such interventions are avoided, or, then, that their occurrence is compensated. In the objective dimension of the fundamental right, there is the state's obligation to create procedural regulations for the guarantee of personal data protection. Here, the provision of procedural mechanisms can be understood as related to the rights of the holders in acts of public power that carry out interventions in the sphere of data protection.

After a decision by the Federal Supreme Court, Constitutional Amendment No. 115/2022, approved by Congress, inserted in Art. 5° the clause LXXIX, with the following wording: "it is ensured, under the terms of the law, the right to personal data protection, including in digital media."

Andorno and Ienca (2017) argue that the current conception of privacy and its dimensions, especially that of personal data protection, are focused on safeguards for external information about people, while neuroprivacy reaches the very source of the production of neural activity as a chronologically antecedent level to the production of personal data.

In Brazil, there were[75] two bill projects that seek to precisely modify the Brazilian General Data Protection Law for the inclusion of neural data: PL 1229/21 and PL 522/22.

Both projects were archived due to the congressman's withdrawal, but started from the premise established in the article "*Your Neural Data Part of Your Mind? Exploring the Conceptual Basis of Mental Privacy*" by Abel Wajnerman Paz, who states that neural data are "*personal information about neural states, processes and structures*," which make up psychological integrity, or the control that individuals have over consciously altering their mental processes, a topic already studied in Sect. 2.1.

---

[74] Mendes (2014), p. 175.

[75] The PLs were filed by request for withdrawal in 2022 by the deputy who presented them (Author's Note).

Thus, neural data would be a category of data that has an ontological relationship with the nervous system and that, therefore, would require a different characterization from that given to sensitive personal data.

Just like personal data protection, neuroprivacy can be understood as brought about by the need for new protections for human dignity in the scenario of mass use of neurotechnologies for entertainment.

References

Andorno R, Ienca M (2017). Towards new human rights in the age of neuroscience and neurotechnology. Life Sci Soc Policy 13:5. https://doi.org/10.1186/s40504-017-0050-1

Bublitz JC (2013) Cognitive Enhancement An Interdisciplinary Perspective My Mind Is Mine!? Cognitive Liberty as a Legal Concept Springer Netherlands Dordrecht 233–264

Celeste E (2019) Digital constitutionalism: a new systematic theorisation. Int Rev Law Comput Technol 33(1):76–99

de Assis Zanini LE, Queiroz ON (2021) A inviolabilidade da pessoa humana e o direito geral da personalidade. Revista Jurídica da Seção Judiciária de Pernambuco 1(13):193–216

De Gregorio G (2021) The rise of digital constitutionalism in the European Union. Int J Constitutional Law 19(1):41–70

Doneda D (2019) Da Privacidade À Proteção De Dados Pessoais, 3ª Ed. Thompson Reuters São Paulo

Doneda D, Zanatta RA (2022) Personality rights in Brazilian Data Protection Law: a historical perspective. In: Albers M, Sarlet IW (eds) Personality and data protection rights on the Internet, Ius gentium: comparative perspectives on law and justice, vol 96. Springer, Cham. https://doi.org/10.1007/978-3-030-90331-2_3

Hoffmann-Riem W (2015) Innovaciones en la Jurisprudencia del Tribunal Constitucional Alemán, a Propósito de la Garantía de los Derechos Fundamentales en Respuesta a los Cambios que Conducen a la Sociedad de la Información. Direito Público 12(64)

Hoffmann-Riem W (2021) A proteção jurídica fundamental da confidencialidade e da integridade dos sistemas técnicos de informação de uso próprio. Direito Público 18(100)

Mendes GF (2013) A dignidade da pessoa humana na Constituição Federal de 1988 e sua aplicação pelo Supremo Tribunal Federal. Observatório da Jurisdição Constitucional 6(2):83–97

Mendes LS (2014) Srivacidade, proteção de dados e defesa do consumidor: linhas gerais de um novo direito fundamental. Editora Saraiva, São Paulo, p 27

Mendes LS (2020) Autodeterminação informativa: a história de um conceito. Pensar Revista de Ciências Júridicas Universidade de Fortaleza (Unifor), Fortaleza 25(4):1–18

Mendes GF, Fernandes VO (2020) Constitucionalismo digital e jurisdição constitucional: uma agenda de pesquisa para o caso brasileiro. Revista Justiça do Direito 34(2):6–51

Pereira JR, Keller CI (2022) Digital constitutionalism: contradictions of a loose concept. Revista Direito e Práxis 13:2648–2689

Sarlet IW (2006) Dignidade da Pessoa Humana e Direitos Fundamentais na Constituição de 1988, 4a ed. Livraria do Advogado, Porto Alegre

Sarlet IW (2020) Dignidade da pessoa humana na Jurisprudência do Supremo Tribunal Federal. Revista da Faculdade de Direito 41(2):15–46

Sarlet IW (2023) Direitos Fundamentais em Espécie. In: Mitidiero DF, Marinoni LGB, Sarlet IW (eds) Curso de direito constitucional, 12th edn. Grupo GEN. https://integrada.minhabiblioteca.com.br/books/9786553624771

Sarmento D (2016) Dignidade da pessoa humana: conteúdo, trajetórias e metodologia. Editora Fórum

Schreiber A (2014) Direitos da Personalidade, 3a ed. Atlas, São Paulo
Supremo Tribunal Federal do Brasil (STF, 2010) Ação Direta de Inconstitucionalidade 3510
Supremo Tribunal Federal do Brasil (STF, 2012) Arguição de Descumprimento de Preceito Fundamental 54
Supremo Tribunal Federal do Brasil (STF, 2020) Ações Diretas de Inconstitucionalidade 6.387, 6.388, 6.389, 6.390
Supremo Tribunal Federal do Brasil (STF, 2022) Ação Direta de Inconstitucionalidade 6.649
Waldron J (2007) Dignity and rank: in memory of Gregory Vlastos (1907–1991). Eur J Sociol/Archives Européennes de Sociologie 48(2):201–237

Chapter 5
Current State and the Future of Neurorights

Internationally, several initiatives have been launched to address the ethical and legal challenges posed by neurotechnologies. The United Nations has played a significant role in this area. For example, on March 12, 2020, Dr. Rafael Yuste, president and founder of the Neurorights Foundation, met with the UN Secretary-General, Antonio Guterres, to discuss the importance of neurorights. This was followed by the presentation of a proposal on neurotechnology and human rights by the Advisory Committee to the UN Human Rights Council on September 8, 2022.

The UNESCO's International Conference on Neurotechnology Ethics, held in Paris on July 13, 2023, and the keynote speech on digital rights delivered by Rafael Yuste at the UNAOC and ONUART Global Dialogue conference in Geneva, on October 9, 2023, further emphasize the growing recognition of neuron rights at the international level. These events highlight the collaborative efforts of various stakeholders to ensure that neurotechnological advances align with fundamental human rights.

In Latin America, the Organization of American States (OAS), through its Inter-American Juridical Committee (CJI), approved in 2021 a Declaration on "Neuroscience, Neurotechnologies and Human Rights,"[1] highlighting the legal challenges arising from scientific and technological advances in the Americas. The text recognizes that neuroscience and neurotechnologies raise ethical and legal questions about fundamental rights such as human dignity, autonomy, privacy, and freedom of thought. The Organization emphasizes the need for regulations to mitigate the risks of manipulation and ensure the protection of individuals' physical and mental integrity.

The declaration also addresses the importance of ensuring equal access to technologies and combating potential discriminatory practices. Issues of informed consent, mental privacy, protection of neural data, and the risks of non-medical uses of technologies are prominently addressed. The CJI recommends that states, the

[1] Organization of American States (OAS, 2023).

A. A. Possa, *Neurorights and the Era of Neurotechnologies*, SpringerBriefs in Law, https://doi.org/10.1007/978-3-031-92002-8_5

private sector, academia, and the scientific community engage in measures to ensure the ethical and safe use of these innovations, promoting public policies that consider human rights.

Finally, the Commission reiterates the need to harmonize national and international standards on neurotechnologies and calls for the participation of various sectors to ensure that such advances benefit people without compromising their fundamental rights.

In the European Union, discussions are being guided by the European Commission, through the Bioethics Committee, and within the framework of the Strategic Plan for Human Rights and Technologies in Biomedicine.[2] In 2021, the document "Common human rights challenges raised by different applications of neurotechnologies in the biomedical fields" was presented, written by Marcello Ienca at the request of the Committee,[3] addressing how neurotechnologies can record and modify neuronal activity, offering significant benefits, such as the treatment of neurological diseases, but also raising ethical and legal concerns. Issues such as mental privacy, personal integrity, autonomy, and protection against misuse of these technologies are discussed.

The overall aim of the document is to provide a critical and normative analysis of the ethical, legal, and social challenges associated with the use of neurotechnologies, identifying priority areas for reflection and regulation, to promote people's well-being and protect their dignity and fundamental rights.

Considering the panorama of neurotechnologies, new risks that may derive from them, the content of possible new rights, and an analysis of the constitutional protections that seek to guarantee human dignity, it is possible to conclude about the problem regarding the need for new fundamental rights or the restructuring of some of these rights in Brazil.

It is important to note that the topic is complex and multidisciplinary, as knowledge from areas such as medicine, biology, neurology, psychology, systems development, law, and even philosophy are needed for effective regulation of the topic.

Thus, it was not proposed and it is not intended to present a regulatory model in the mold of those studied in Chap. 3. The work may assist in the academic and legal discussions that are being carried out in Brazil on the issue, with a study on the concretizations of the protection of human dignity in the form of fundamental rights.

These studies that seek to substantiate the need to create rights with international and constitutional status are based on the premises that new types of technologies that interact with brain activity will be more accessible to the mass public, with various possibilities of use beyond those currently present in the medical field, resulting in new worrying risks about human self-determination.

The first point to highlight in the discussions is this premise related to risks.

The risks that are normally presented are results of academic and laboratory research, with prototypes that, in many cases, probably did not even continue to be developed, considering the time lapse since the cited publications.

[2] European Commission (EC, 2019).

[3] Ienca (2021).

These risks, therefore, in addition to having a certain abstract character, considering they originate from technologies that may not reach a considerable range to justify new rights, may never come to be realized. Refining this point, it is necessary to mention an observation made in relation to the characterization of devices as neurotechnologies: many of them, in reality, probably respond to the electrical stimuli performed by the muscles that are in direct contact with the device, as opposed to true neurotechnologies that interact directly with these stimuli.[4]

The validity of this premise is extremely important for legal constructions that can create obligations for technologies that only apparently interact with neural data, creating burdens for already highly regulated areas.

In the case of neurotechnologies, the risks are still at an abstract level due to the insufficiency of cases to be observed in practice, since there are no BCIs marketed on a large scale in any economic segment.

The problem of abstract risks for the structuring of new rights can result in normative constructions without a scope of protection that corresponds to the practical problems that neurotechnologies can end up creating, creating a protective vacuum for the concrete damages that exist.

Therefore, the research has the utility of indicating a possible path that can be considered taking into account the fragility of the premises and the need for normative support already accepted for the system of fundamental rights in Brazil: the identification of neurorights as manifestations of the protection of human dignity against risks of neurotechnologies.

The analysis of the proposals for neurorights in Chap. 2 and the comparison of them with the recognized personality rights (understood here as constitutional realizations of the protection of human dignity, and not in their scope of private law regulation) allow some conclusions that can assist future studies.

Considering the problem highlighted above about the premises on the risks of neurotechnologies, in the current scenario, a comprehensive and comprehensive protection model does not seem viable, and, perhaps, not even desirable.

Therefore, among the two sets of neurorights (the proposal of Rafael Yuste and Sara Goering in the magazine *Nature* and the proposal of Ienca and Andorno), the right to cognitive freedom manifests itself as a fundamental right structured sufficiently to be considered a new right.

As highlighted, cognitive freedom stands as a right that protects a dimension of general freedom for the guarantee of individual self-determination: the possibility of choosing to use or not to use neurotechnologies with the security of neural computations, both at levels of biological structure and behavior, against actions of third parties that may cause loss of bodily or psychological self-determination and damages by third parties (both those that can be malicious, as well as those that can derive from development failures).

It emerges as a general protection clause against the risks derived from devices that interact directly with brain activity, regardless of other specific protections that may be considered in the future for new dimensions of legal protection (like the case

[4] Rainey et al. (2020), p. 2300.

of neuroprivacy, which was indicated as a possible new dimension of privacy, but still backed by risk premises that require further empirical studies).

In the same way that the right to free development of personality is autonomous and guarantees protection for human dignity in cases of risks that are not protected by specific rights, cognitive freedom can fulfill a similar function as the existence of damages resulting from neurotechnologies becomes part of social reality.

The dimensions of Bublitz JC (2013) for this right that were studied in Sect. 3.3.1 guarantee, even, the theoretical foundation for the future structuring of species of neurorights as realizations of the protection of human dignity, which can bring tranquility to the researchers who seek a regulatory model based on the apprehension of repetition of problems that are currently existing due to the inertia of the legal plan in front of the large social platforms.

The rights to mental integrity and psychological continuity, apparently, would already be fully contemplated in the rights to physical and psychic integrity identified by Ingo Sarlet, not demanding legal innovations.

The neuroright to neuroprivacy (or neural privacy) can be understood as a new dimension of the right to privacy or even a new autonomous fundamental right not limited to it, in the same way that Danilo Doneda understands the right to protection of personal data. However, considering the problem of the lack of concrete predictions about the risks and damages of neurotechnologies, it is not possible to glimpse, at the present moment, elements suitable for a normative construction.

Thus, the present work demonstrates a legal reasoning based on the international and national context of the protection of the dignity of the person human that can be an instrument in the construction of the right to cognitive freedom in the constitutional order of Brazil and can, even, serve as method for construction in the scope of human rights or other constitutional orders.

References

Bublitz JC (2013) Cognitive Enhancement An Interdisciplinary Perspective My Mind Is Mine!? Cognitive Liberty as a Legal Concept Springer Netherlands Dordrecht 233–264

Comissão Europeia (2019) Strategic action plan on human rights and technologies in biomedicine. https://rm.coe.int/strategic-action-plan-final-e/1680a2c5d2. Accessed 19 Nov 2024

Ienca M (2021) Common human rights challenges raised by different applications of neurotechnologies in the biomedical field. Report commissioned by the Council of Europe

Organização dos Estados Americanos (2023) Declaración De Principios Interamericanos En Materia De Neurociencias, Neurotecnologías Y Derechos Humanos. https://www.oas.org/es/sla/cji/docs/CJI-RES_281_CII-O-23_corr1_ESP.pdf. Accessed 19 Nov 2024

Rainey S et al (2020) Brain recording, mind-reading, and neurotechnology: ethical issues from consumer devices to brain-based speech decoding. Sci Eng Ethics 26(4):2295–2311. https://doi.org/10.1007/s11948-020-00218-0

Made in the USA
Monee, IL
14 September 2026

62974048R00046